Classic Supercars

Classic Supercars

RICHARD GUNN

amber
BOOKS

This edition first published in 2017

Reprinted in 2023

Published by Amber Books Ltd
United House
North Road
London
N7 9DP
United Kingdom
www.amberbooks.co.uk
Instagram: amberbooksltd
Facebook: amberbooks
Twitter: @amberbooks
Pinterest: amberbooksltd

ISBN: 978-1-83886-330-2

Project Editor: Michael Spilling
Design: Hawes Design

All pictures © International Masters Publishing AB

Printed in China

Contents

INTRODUCTION

Here is a surprising fact to start off with: the word 'supercar' does not appear in most dictionaries. And although the term has been in common use for decades now, its true meaning is still hard to pin down. Even the normally definitive *Oxford English Dictionary* affords it only the briefest of descriptions. While this traditional vanguard of vocabulary goes into great detail on all manner of other well-known and obscure words, on the subject matter of this book, it is inadequately vague. A supercar is, apparently, a '...high-performance sports car.'

Well, yes, that is true, to a certain extent. A supercar is a high-performance sports car – but also so much more than that. There are many sports cars around these days, most of them with the sort of looks, performance and handling to humble even the greatest machines of yesterday. But do these mass-produced modern wonders qualify as supercars? The answer is a resounding no.

A supercar is something very, very special, the cream of the motoring world, the type of car that ordinary mortals lust after, but which few will ever even drive, let alone own. They are the stuff of teenagers' poster collections, millionaires' driveways and everybody elses' dreams. To meet the 'super' criteria, a car has to be spectacular, whether standing still or scything through the air at high speed. Step forward Ferrari, Lamborghini, Porsche, Aston Martin and very few others.

Supercars are exclusive machines. You will not see them parked on the street corner or doing the shopping run to the local mall. Some are now so rare or valuable that the only place to see them is at car shows or museums, while a very few are never even

seen by the public, just playthings of the rich for whom the pleasure of owning a motoring icon outweighs their need to show it off.

A supercar need not necessarily be beautiful, but it does have to be eye-catching. The majority of such machines are indeed attractive, their need to be aerodynamically effective dictating gorgeous flowing lines and a smooth, wind-cheating shape. But others border on the downright unattractive,

When it comes to supercars, Ferrari eclipses all others. The F40 was an attempt to build the ultimate supercar – and it came very close to that ideal.

all bulging wheel arches, greedy air scoops or aggressive angular lines. Take, for example, the Lamborghini Diablo. It is not a pretty car, but it stands out from the crowd like nothing else. And while the Bristol Beaufighter has all

7

the visual appeal of a breeze block on wheels, it is this very aspect that most endears it to enthusiasts. Both are distinctive and both are celebrated in this book.

There is usually something about the way a supercar sounds as well. Dramatic looks generally go hand in hand with a wonderful noise from the exhausts. How can anybody with the remotest interest in cars fail to be moved by the commotion of an Aston Martin engine bursting into life with a throaty roar, or an ear-splitting Lamborghini V12 screaming through its revs? Either at rest or in full flight, most supercars deliver the kind of soundtrack that elevates their engineers to the standing of composers. Often, you will hear a supercar approaching long before it is visible, but just the mechanical fanfare is enough to make you stop, turn around and strain to catch a glimpse of what is causing all the uproar.

All this would matter little, though, without substance, and away from the adoring fans, a proper supercar must offer exhilarating performance as well. That means more than just speed in a straight line, although obviously, the faster a car is, the more likely it is to have supercar status bestowed on it. But sheer velocity should also be accompanied by road-holding and braking ability as well. For some of the earlier examples included in this book – such as the Porsche 356, Alfa Romeo 8C 2900 or SS100 – even reaching 160km/h (100mph) was a challenge. Yet for their era, they offered a beguiling driving experience, blending a combination of engineering factors that made them far superior to their run-of-the-mill contemporaries. And even though these cars have long since been dynamically superseded by younger upstarts, they are still cherished as greats. Is a Bentley 4.5l, with a top speed of 200km/h (125mph), any less desirable than a current Ferrari 360 Modena capable of over 320km/h (200mph)? Some enthusiasts swear by the Prancing Horse badge, while others refuse even to consider anything without a vintage Flying B on its radiator grille. The two may feel completely different out on the road or track, but what they share in common is this: when new, they were just the best around.

So when were the first supercars, the first extraordinary vehicles, born? Once upon a time, of course, there was no such thing. There were only automobiles, those noisy, wheezing and unreliable veterans that were the preserve of the affluent and eccentric. Back in those pioneering days of motoring, when even getting an internal combustion engine just to run properly for a reasonable length of time was challenge enough, few bothered about speed. And in Britain, the first cars were required to have a man

walking in front of them waving a red flag and were confined to speeds of 6.5km/h (4mph), hardly the type of environment to nourish thoughts about increasing performance. In a way, it is justifiable to call all of these early vehicles supercars, as they were so novel, so technologically advanced, that they were regarded as wonders of the age. Whenever one clattered into view, bystanders could look at nothing else.

But, as mechanical trustworthiness improved and motoring became more widespread, so some cars started to develop

American supercars are less numerous than their European counterparts, but the original V8 Chevy Corvette fitted the description superbly.

in other directions aside from just technical competency. Styling and design became more significant, and speed became a major selling factor. At first, attempts at boosting performance were simplistic, with just a little extra power here, or a touch of body lightening there. But soon, increasing engineering sophistication, such as supercharging and the use of aviation principles, took over. By the 1920s, sports cars had become established as an automotive race apart, and within this pedigree, there were those regarded as the absolute elite. If you wanted something cheap and cheerful that did its job and little else, you bought an Austin or a Ford. If you wanted something special that would be just as competent on a race track as it was on ordinary roads, then badges such as Bentley or Alfa Romeo were among the choices. Arguably, the French firm of Bugatti was the first proper supercar marque, building highly specialized, expensive, fast and glamorous

machines that could be worked on only by experts. A Bugatti was the type of car you could never hope to own unless you were prepared to put up with its eccentricities. And to be able to afford to do so, you had to be very, very rich…

It was only after World War II that the first acknowledged supercar was created, although the term would later be applied retrospectively. In 1954, Mercedes-Benz unveiled the magnificent 300SL, which had its origins in racing. The Le Mans-winning coupe was translated into a superlative road car, without equal at that time. Not only did the 300SL look sublime, having gullwing doors and a charismatic aura, but with its space frame chassis, it also featured engineering principles that had not been used on a road car before. And it could reach 242km/h (150mph), a sensational figure for the time.

The 300SL set a precedent. Although it was in production for only a short time and

Aston Martin's glorious DB7 is one of the best of the current breed of supercars, signalling a revival for the classic British sports car marque.

came with a price tag that was more than some people would earn in their lifetime, it proved that a market for such high-priced, complicated supercars existed. That was good news for Italian Enzo Ferrari, the former Alfa Romeo racing mechanic who had started building road cars under his own name in the late 1940s. In the years that followed, Ferrari would become one of the most famous marques around, creating a range of supercars that none could equal. Other manufacturers seeking to imitate

Ferrari would spring up (many of them in Italy, thus giving rise to the country's image as the spiritual home of the supercar), but Ferrari would remain almost untouchable, having only three serious competitors. Porsche, from Germany, built its cars from a different mechanical standpoint, while Lamborghini – a company established after an argument between Enzo Ferrari and tractor builder Ferruccio Lamborghini – set out to make cars intended to outdo its rival in every sense. Indeed, it was Lamborghini's Miura that kick-started the fashion for mid-engined sports cars, becoming the first machine to be called a supercar while still in production. Meanwhile, in England, Aston Martin found itself having to compete against Ferrari, Porsche and Lamborghini

11

The handsome Ghia 450SS was the dream of an American businessman, but the car failed to survive long enough to realize its true potential.

almost by accident, after its 1963 DB5 found fame as the car of James Bond 007, and was suddenly one of the most desirable cars in the world. All of these manufacturers are still around today, still building the kind of cars that motoring journalists run out of superlatives to describe, and mere mortals lust after.

What of the future? The death of the breed has been predicted for decades. After the optimistic years of the 1960s, when supercars seemed to be launched at the rate of one a week, the energy conscious, recession-hit era of the 1970s came as a bitter shock. As the world lurched from one fuel crisis to the next, expensive and thirsty big-engined monsters suddenly found themselves out of favour.

However, the storm was ridden out, and when life returned to something resembling normality, they were still with us. But then came safety legislation and environmental concerns. How could anything capable of breaking speed limits several times over possibly be safe? Why should exotic gas-guzzlers be allowed to continue sapping the Earth's resources? Such issues presented a long-term problem for the car makers. Their solution was to adapt, making engines cleaner and more efficient, building in extra protection for occupants and pedestrians, while at the same time trying not to compromise the performance that supercars have always provided. On the whole, the car firms are winning the battle. Recent Ferraris now have lower emissions than ever before, and are safer in every way. But they can now reach over 320km/h (200mph) almost as a matter of course, and thanks to the use of electronics and better technology, have the kind of handling that was simply impossible even a few years ago.

Whatever the future holds, supercars will always be with us. There will always be car firms who want to build something bigger, better and faster than their rivals. The world needs spectacular cars, even if it is just to balance out all the dull and ordinary ones.

Until the next supercar comes along to stagger the world – and no doubt the latest one is just around the corner, straining to be unleashed – enjoy this book of past and present greats. The 75 presented here are just a small selection of the greatest cars ever made, but we think they are some of the best. Without them, the world of motoring would have been far less fascinating.

AC 428

After the legendary Cobra came the AC 428, an altogether more civilized creature than its venom-spitting predecessor, made in Britain with American power and Italian styling.

The Italian design house Frua was responsible for styling and building bodies for the AC 428. Unfortunately, their design was not new: the look had first appeared on the Maserati Mistral of 1963. There are a few detail differences, but put a Mistral besides a 428 and you will see a strong resemblance.

The 428 of the AC's title referred to its engine, a big-block Ford V8, made from iron and used in some of the biggest US Fords of the era. It was a more flexible unit than its predecessor, the 427, and better suited to the AC's long-limbed grand touring ability.

Although it looked very different on top, underneath the 428 was a modified tubular steel ladder frame Cobra chassis, with lengthened chassis rails.

Inside was one of the most comprehensively-equipped dashboards around, with a large bank of rocker switches occupying the vast centre console. In front of the driver, eight gauges monitored what the car was doing.

You could never specify an automatic gearbox for the Cobra, but the 428 came with one as standard, a Ford C6 three-speed unit.

The British company AC Cars Ltd probably had every right to feel a bit conned when the marque's newest car came out in 1966. It was a fine-looking machine, with polished styling and a European chic to it that few other British cars possessed at the time. There was one slight problem, though. It looked almost exactly like the Maserati Mistral, a design that had appeared three years earlier. Low-volume manufacturer AC, without the resources to create a body in-house for its

The Ford 428 V8 engine in the AC generated most of its power and torque low down in its rev range, making the car more suited to relaxed cruising than outright speed.

new car, had turned to Italian coachbuilder Frua to do the job. In effect, Frua had sold them an existing design.

This aside, the AC 428 was a smart entry into the supercar world of the late 1960s. It was built as a replacement for the Cobra, but was actually more of a complete change of direction, moving away from sports car territory and into the world of grand tourers. It offered far more refinement than the Cobra, although this did come with a higher price tag.

WELL-DISGUISED COBRA

The 428 used the Cobra's robust chassis, and lengthened it, modifying the wishbone suspension at the same time. Onto this, the Frua body was fitted – once it had arrived from Italy – and a Ford 428 V8 engine was installed under the hood. The interior was luxuriously trimmed out, with a healthy complement of controls and instruments to keep the driver happy.

The 428 was always an exclusive model, and just 29 coupes and 51 convertibles were made before industrial problems and the fuel crisis forced AC to push it aside in 1973.

AC 428

Top speed:	228.5km/h (142mph)
0–96km/h (0–60mph):	6.4 secs
Engine type:	V8
Displacement:	7016cc (428ci)
Max power:	257kW (345bhp) @ 4600rpm
Max torque:	626Nm (426lb-ft) @ 2800rpm
Weight:	1461kg (3214lbs)
Economy:	2.83km/l (8mpg)
Transmission:	Ford C6 three-speed automatic
Brakes:	Four-wheel Girling discs
Body/chassis:	Tubular-steel ladder frame with welded-on steel two-door convertible or fastback body

UNITED KINGDOM

UNITED KINGDOM

AC Cobra Mk IV

The epic AC Cobra died in 1968. Or did it? Proving that it is impossible to keep an immortal motoring legend down, AC reincarnated it in 1985.

Following tradition, a Ford V8 also powered the 'new' Cobra, although it was a modern 5-litre (305ci) unit as used in Mustangs of the era. Originally, it was fitted with a four-barrel Holley carburettor, but this was superseded by electronic fuel injection in 1985.

Just like a traditional Cobra, it was more than easy to scrape the exhaust when tackling bumpy roads. Side exiting pipes were an option, which increased both power and the awesome soundtrack of the V8.

The cockpit was bigger inside than on the 1960s Cobras. And instead of the metal dash, owners got one trimmed in leather instead. However, for those who chose the special Mk IV Cobra Lightweight, the dashboard was finished in metal. These Lightweights also had more powerful carburetted engines.

Although it looked exactly like the last of the original Cobras from the 1960s, the Mk IV was subtly changed. It was both wider and longer, although the wheelbase stayed the same. One of the reasons it was broader was because the wheel arches had to be flared to cope with fatter modern tyres.

UNITED KINGDOM

No sports car had come along to equal the AC Cobra following its premature demise in 1968. Sure, there had been faster vehicles, more powerful ones and others that cornered more quickly. But nothing had quite captured the same level of red-blooded, raw excitement as Carroll Shelby's original open two-seater of the 1960s.

So, AC decided to bring it back, for a new generation that had never experienced the real thing. This was not the original AC

There was never any doubt about what would power the reborn Cobra. It had to be a Mustang V8, just as with the original legend. However, the Mk IV used a modern incarnation, as also used in contemporary Mustang GTs and LXs.

firm, but a British Cobra specialist called Autokraft. It bought the rights to the AC name, as well as all the Cobra tooling and jigs in 1982, and in 1985 unleashed the AC Cobra Mk IV on an unsuspecting world.

REVIVAL OF THE FITTEST

It did not quite have the same calibre as the original, but it came close. And considering all the restrictive legislation that had come along since the 1960s, it was still impressive. Featuring – what else? – a Ford V8 from the modern Mustang, around 480 Cobra Mk IVs were built until the mid-1990s. Despite some superficial cosmetic changes, the reborn Cobra managed to project the aggressive machismo of the bona fide classic, particularly the stripped out Lightweight model, which came closest to the spirit of the old Cobra 427. The new incarnation was not quite as fast or powerful, but its handling was better, and driving it was less of a constant battle between man and machine.

In 1997, the supercharged Superblower ousted the Mk IV, offering 264kW (355bhp), even more than the 427 had boasted.

AC Cobra Mk IV

Top speed:	217km/h (135mph)
0–96km/h (0–60mph):	5.3 secs
Engine type:	V8
Displacement:	5000cc (305ci)
Max power:	168kW (225bhp) @ 4200rpm
Max torque:	406.5Nm (300lb-ft) @ 3200rpm
Weight:	1122kg (2469lbs)
Economy:	6.02km/l (17mpg)
Transmission:	Borg-Warner T5 five-speed manual
Brakes:	Four-wheel vented discs
Body/chassis:	Tubular steel chassis with cross members with alloy two-seater convertible body

UNITED KINGDOM

21

ALFA ROMEO 8C 2900

Supercars before the term was first used, Alfa Romeo 8Cs were effectively full-blown racing cars clad in some of the most artistic bodies ever created.

Alfa Romeo did not generally construct the bodies for the 8C series, but usually just supplied the running chassis for specialist coachbuilders to build upon. However, this body was built by Alfa, on a shortened chassis, and was far less exuberant than those made by other builders.

An 8C engine could get very hot under the hood, so most of the bodies had vents down the side of the hood doors. On this model, some of the flaps were adjustable.

8C models came either as long chassis (Lungo) or short chassis (Corto) versions.

Alfa understood the importance of balance in a sports car, so the four-speed transmission was mounted at the rear of the car.

Twin superchargers were fitted to the straight-eight cylinder engine, which was of an unusual design, with two separate twin-cam four-cylinder blocks driven off a central crankshaft. The heads and the blocks were made as one item, so there was no need for gaskets as on a more conventional engine.

Crude independent rear suspension, using swing axles, was fitted on the 8C.

ITALY

23

I talian engineer Vittorio Jano came to Alfa Romeo in 1923 and started a new and successful chapter in the prewar history of the Milanese marque. Under his guidance, Alfa was dominating racing by the end of the decade, thanks to its P2 Grand Prix cars.

If Alfa was to stay ahead, it needed to continue to develop its racing technology. Jano started work on a supercharged straight-eight twin-cam engine, which would go on to become the staple power source for 8C series of cars. It was a

The two camshafts were driven by a central drive mounted between the two cylinder banks of the 8C's straight-eight. This novel arrangement got around the possible problem of camshaft flexing.

complex device, made out of two separate four-cylinder blocks, and its monobloc construction meant that head gaskets – often unreliable in engines of this era – were unnecessary. It would later be recognized as one of the greatest engines of the era.

FERRARI'S INSPIRATION?

The first car the engine appeared in was the 8C 2300 of 1931. But it was in 1935 that the most advanced and powerful development appeared, as the 8C 2900. The cars were used for racing, and also sold privately to the well-heeled gentry of Europe.

As befitted the advanced engineering, some gorgeous bodies were sculptured by coachbuilders such as Zagato, Touring and Pininfarina. Each company, it seemed, was trying to outdo the others to produce the most exotic, art deco creations possible.

Extremely successful in racing, and extraordinary on the road, the 8C's production was cut short by World War II.

There was a footnote, though: an engineer who had prepared 8Cs for Alfa's racing programme went on to form his own marque after the war ended. His name was Enzo Ferrari…

Alfa Romeo 8C 2900B

Top speed:	185km/h (115mph)
0–96km/h (0–60mph):	9.6 secs
Engine type:	Supercharged in-line eight
Displacement:	2905cc (177ci)
Max power:	134kW (180bhp) @ 5200rpm
Max torque:	Not quoted
Weight:	1145kg (2519lbs)
Economy:	Not available
Transmission:	Rear-mounted four-speed manual
Brakes:	Four-wheel drums
Body/chassis:	Various coach-built styles on steel box section ladder frame with cross members

ITALY

ALFA ROMEO MONTREAL

As Alfa Romeo's flagship of the 1980s, the Montreal was the one of the marque's most exciting cars, with a race-bred engine and novel styling.

These fake vents down the side of the Montreal made it look like it was mid-engined, but the engine was conventionally installed in the front. They also meant that rear seat passengers had very little visibility. The embellishments were left over from the Montreal's first appearance as a 1967 concept car.

The all-alloy V8 engine of the Montreal came from the company's Tipo 33 racing car, although it was detuned for road use. It had four camshafts (two on each cylinder head), and received its fuel via a Spica mechanical fuel injection system.

As unusual as the Montreal seemed on the surface, it was conventional underneath, based on Alfa Romeo's 1750/2000 Berlina series. With the Montreal's surfeit of power, this meant that handling could get choppy in certain circumstances. That said, the four-wheel vented discs helped keep less competent drivers out of trouble.

Quad-headlamps had small grilles above them. This was purely a stylistic, although distinctive, feature, which made the headlamps inconspicuous when not switched on.

The Montreal was available in left-hand drive only, and just as a fixed-head coupe.

ITALY

A t the 1967 Expo show in Montreal, Canada, Alfa Romeo had something special to show the world. On its stand was a new show car, a radical-looking coupe design by Bertone. Its unorthodox appearance - with slats decorating its rear roof pillar and vents obscuring the headlights - drew great interest from both press and public alike, but nobody expected anything much to come of it. After all, concept cars never actually go into production, do they?

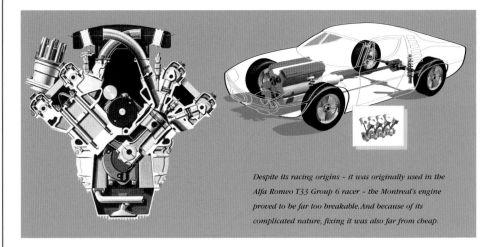

Despite its racing origins - it was originally used in the Alfa Romeo T33 Group 6 racer - the Montreal's engine proved to be far too breakable. And because of its complicated nature, fixing it was also far from cheap.

This one did, however. Impressed by the reaction to the car, Alfa Romeo prepared it for manufacture, and revealed it again in 1970, this time as a version the public could purchase. And this time, its engine was just as tempting as its looks. Installed under the hood was a detuned version of Alfa's Tipo 33 V8 racing engine.

Although the car's visual aspect hinted at a mid-engine configuration, with its thick rear roof pillars, glass rear hatch and those promising vents, the engine was at the front. It was mounted on an Alfa Romeo 1750 sedan chassis.

FUEL CRISIS FAILURE

Named Montreal after the city where it had first been glimpsed, it was the fastest road Alfa ever produced, but unfortunately, this Alfa was another 1970s case of right car, wrong time. Its V8 engine was a thirsty beast during a fuel-conscious decade (despite being fitted with mechanical fuel injection),

and the highly specialized unit also developed a reputation for fragility. Less than 4000 managed to find homes before it was replaced by the GTV6 in 1977, a less spectacular but more conventional and reliable sports coupe.

Alfa Romeo Montreal

Top speed:	220km/h (137mph)
0–96km/h (0–60mph):	8.5 secs
Engine type:	V8
Displacement:	2593cc (158ci)
Max power:	149kW (200bhp) @ 6400rpm
Max torque:	237Nm (175lb-ft) @ 4750rpm
Weight:	1278kg (2811lbs)
Economy:	6.37km/l (18mpg)
Transmission:	Five-speed manual
Brakes:	Four-wheel vented discs
Body/chassis:	Monocoque chassis with steel two-door coupe body

ITALY

29

ASTON MARTIN DB4 ZAGATO

The DB4GT was a stunning enough Aston in its own right, but a lithe new lightweight body from Zagato turned it into an electrifying performer.

Weight was considerably less than the standard DB4, thanks to the shorter wheelbase (13cm/5in less than a normal DB4), light alloy bodywork, and the substitution of Perspex for glass in the side and rear windows.

The Zagato body bore a resemblance to the standard DB4, but had smoother, more rounded ends to improve air flow. The headlamps were covered with plastic to aid the aerodynamics as well. Only one car was built with exposed headlamps.

There were – count them – 72 spokes in a Zagato wire wheel. This number – more than on a usual wire wheel – was needed to stand the stresses caused by the extra power and torque of the uprated engine.

Beneath the hood was the trusty DB six-cylinder engine, used in all of the company's cars during the 1960s. For these GT cars, though, it was modified to provide the sort of power its looks demanded.

Quick-release fuel caps were fitted on both rear fenders, a legacy of the Zagato's use in racing.

Another weight-saving technique was the omission of bumpers.

The DB4 came out in 1958, the first in a visually similar line of DB models that would endure until the DB6. A year later came a higher-performance version, the DB4GT, which was shortened, lightened and mechanically uprated.

More was to come: inspired by a Zagato-bodied Bristol 406, the general manager of Aston Martin met with Italian coachbuilder Gianni Zagato to discuss the possibility of creating the ultimate DB4, using bodywork designed and built by the Milanese company.

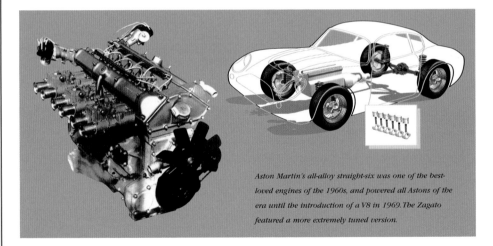

Aston Martin's all-alloy straight-six was one of the best-loved engines of the 1960s, and powered all Astons of the era until the introduction of a V8 in 1969. The Zagato featured a more extremely tuned version.

It took just a year for the first car to appear, unveiled at the 1960 London Motor Show. This was a beautiful creation, the work of Zagato's young designer Ercole Spada. Because of its hand-built nature, and the need to accommodate customers' specific demands, each Zagato was different from its sisters, but they all shared Spada's basic design. Amazingly, it was one of his first.

PRICED FOR PERFORMANCE

Fitted with a version of the DB4GT engine with two spark plugs per cylinder, the Zagato was able to exceed 241km/h (150mph), which made it among the fastest cars available at the time.

The Zagato was an incredibly expensive car, a factor which contributed to its low production run of just 19 cars. In the eyes of many, of course, that made it even more special and desirable.

Production 'officially' ended in 1963, although in 1991 four special Sanction II DB4GT Zagatos were built, using chassis numbers unused – or unsold – during the 1960s. These new classics were constructed with the full approval of Aston Martin, and with the assistance of Zagato. They were all but identical to the original series.

Aston Martin DB4 Zagato

Top speed:	244.5km/h (152mph)
0–96km/h (0–60mph):	6.1 secs
Engine type:	In-line six
Displacement:	3670cc (326ci)
Max power:	234kW (314bhp) @ 6000rpm
Max torque:	377Nm (278lb-ft) @ 5400rpm
Weight:	1257kg (2765lbs)
Economy:	4.18km/l (11.8mpg)
Transmission:	Four-speed automatic
Brakes:	Four-wheel vented discs
Body/chassis:	Separate box-section chassis with alloy two-door coupe body

ASTON MARTIN ZAGATO

Aston Martin's links with Italian coachbuilder Zagato stretched back to the 1960s, but were revived again in the 1980s for a new and different-looking supercar.

Some enthusiasts considered the all-aluminium Zagato shape to be too ordinary for an Aston Martin, and it was less eye-catching than previous Zagato efforts. However, it was totally contrary in look to the standard V8 on which chassis it was based.

The engine was a modified version of Aston Martin's long-lived, all-alloy V8. For the Zagato, it featured bigger intake and exhaust ports, improved exhaust manifolds, tweaked twin-overhead camshafts and a higher compression ratio. The pistons were Cosworth items, and fuel was delivered via four Weber carburettors or fuel injection.

Most convertibles had flat hoods as they were usually fuel-injected. Coupes had a hood bulge to clear their carburettors.

This car is the rarest variant of the Zagato, the Volante (a term used by Aston Martin to denote a convertible model). When built, it was the fastest production convertible available. Nearly twice as many hardtop coupes were built.

Apart from the mechanical changes, other ways of improving top speed were a more aerodynamically efficient body, as well as foam-filled bumpers in place of the usual metal ones.

UNITED KINGDOM

In 1984, Aston Martin found itself opposite the Italian coachbuilder Zagato's stand at the Geneva Motor Show. The positioning was a fortunate coincidence for both firms. The two had collaborated back in the 1960s to bring the world the lovely DB4GT Zagato, but had not worked together since.

Looking for a way to refresh Aston Martin's image, chairman Victor Gauntlett realized that a revival of this classic partnership would be mutually beneficial. Thus plans

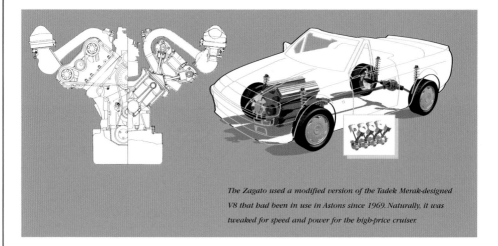

The Zagato used a modified version of the Tadek Merak-designed V8 that had been in use in Astons since 1969. Naturally, it was tweaked for speed and power for the high-price cruiser.

were drawn up for Zagato to build a limited edition model on the V8 chassis, one that not only looked very different but was also endowed with better performance.

The plan was to have a prototype up and running by the 1985 Geneva show. That failed to materialize, but sketches of what the car would look like were there. On the basis of those alone, 50 buyers placed orders worth almost seven million US dollars.

Production of the real thing started the following year, with chassis going out to Italy to be fitted with bodies, then returning to England to be finished. Although the design was less radical and attractive than many had hoped, performance was sizzling, far outstripping that of previous Astons.

TOP CHOP

Attention turned to the Geneva Show once again in 1987, when the convertible version – the Volante – was revealed. Fitted with a less powerful engine, but with fuel injection,

the Volante lacked the large 'power bulge' of the coupe, which had come in for criticism from some of the motoring press. Only 25 of these would be built before production of all Zagatos ceased in 1989.

Aston Martin Zagato

Top speed:	294.5km/h (183mph)
0–96km/h (0–60mph):	4.8 secs
Engine type:	V8
Displacement:	5340cc (326ci)
Max power:	322kW (432bhp) @ 6200rpm
Max torque:	535Nm (395lb-ft) @ 5100rpm
Weight:	1650kg (3630lbs)
Economy:	4.25km/l (12mpg)
Transmission:	Five-speed automatic
Brakes:	CBS linked brakes
Body/chassis:	Steel substructure with alloy two-door coupe body

ASTON MARTIN VIRAGE

The first completely new Aston Martin for 20 years was a big, brutal bruiser of a machine,
uncompromising yet luxurious, expensive and very, very exclusive.

Available alongside the two-door coupe body style
were the Volante (the convertible), a Lagonda-badged
Shooting Brake station wagon, and four-door versions
that also wore the Lagonda name.

Each engine had a plaque on it,
identifying which employee hand-built it.

Underneath the Virage was a variation on the venerable old Aston
Martin chassis used since 1967. However, for the Virage, it was
made lighter and stiffer. The car was a semi-monocoque, with the
aluminium body panels mounted on an inner steel shell.

The Virage used Aston Martin's long-running all-alloy V8 engine, although this was modified by American tuning guru Reeves Callaway to produce more power. The Vantage version would later use twin mechanically driven superchargers to boost bhp even more.

Transmission was a robust ZF unit of German origin. On the standard Virage, there were five forward ratios, but on the more powerful Vantage, a six-speed gearbox was fitted. An automatic transmission was also available, although this took some of the sportiness away.

Rear suspension was unusual for a 'modern' supercar, using an alloy De Dion back axle, with Watts linkage.

A ston Martin had been promising a new car for years. Its V8 model had been in existence since 1969, although the shape dated back to the DBS of 1967. The problem was, there never seemed to be enough money in the coffers to develop anything fresh. And so, the V8 model lingered on, year after year after year.

Finally in 1986, work started on project DP2034, a long overdue replacement for the V8. It had to be a global car, one that could meet the regulations of countries across the

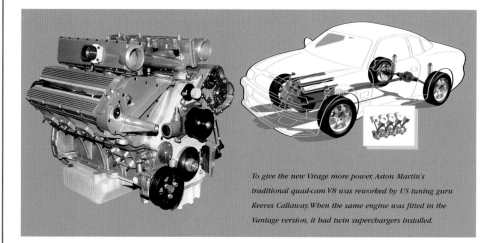

To give the new Virage more power, Aston Martin's traditional quad-cam V8 was reworked by US tuning guru Reeves Callaway. When the same engine was fitted in the Vantage version, it had twin superchargers installed.

world. But most of all, it had to be typically Aston Martin: luxurious, powerful and very, very fast.

An expectant world got its first glimpse of the car at the 1988 British Motor Show. It still used the ancestral V8 engine, and the chassis was carried over from the old car, but both were substantially changed. V8 specialist Reeves Callaway Engineering of the United States was responsible for the engine work. The body was radically different, blending tradition with modernity, thanks to John Heffernan and Ken Greenley, tutors at London's Royal College of Art.

ALL THE VS

Production began in 1989, and by 1990 the first Volante convertible had driven onto the market. The most exciting variant arrived with a blast in 1993. The Virage Vantage pumped out 410kW (550bhp) thanks to its twin superchargers, and had improved suspension as well as muscularly modified

bodywork, making it clear that this supercar meant business.

The Virage title was dropped in 1994, although the cars themselves continued under just the V8 name, but using the Vantage body style as standard.

Aston Martin Virage Vantage

Top speed:	306km/h (190mph)
0–96km/h (0–60mph):	4.6 sec
Engine type:	V8
Displacement:	5340cc (326ci)
Max power:	410kW (550bhp) @ 6500rpm
Max torque:	745Nm (550lb-ft) @ 4000rpm
Weight:	1922kg (4230lbs)
Economy:	4.60km/l (13mpg)
Transmission:	ZF six-speed manual
Brakes:	Four-wheel vented discs
Body/chassis:	Steel skeleton with alloy panels forming two door 2+2 coupe

ASTON MARTIN DB7

Almost a quarter of a century separated the DB6 and DB7, but Aston Martin's 'comeback' sports car was more than worthy of the DB title.

If the Aston Martin DB7 looks like a Jaguar XK8, that is because the two cars share a common heritage. Both firms are owned by Ford, and the DB7 was originally a Jaguar prototype. The XK8 appeared two years after the Aston.

Body design was by Ian Callum, who had worked on the Jaguar XJ220 project. The DB7 uses an alloy and composite body, to keep weight down. Being an Aston Martin, of course, it is completely hand-built.

The cooling vents behind the front wheels, complete with central chrome bar, are something of an Aston Martin trademark, first seen on the DB4 of 1958. Another link with the past is the wheelbase, which is exactly the same as a DB6.

At the heart of the original DB7 was an alloy in-line six-cylinder engine with twin overhead camshafts, developed from a Jaguar XJ40 unit, but fitted with an Eaton supercharger to bestow extra power on the Aston. A V12 engine was a later addition in the Vantage of 1999.

The disc brakes are only vented at the front, but still highly effective.

UNITED KINGDOM

43

The history of Aston Martin has been marked by financial insecurity. Despite building some of the world's most desirable sports cars, the marque struggled to make a profit. All that changed in 1987, when the small British company was bought by global giant Ford. At a stroke, it suddenly found itself with the budget to develop new models, and access to Ford's technical resources in order to do so.

Under the new chairmanship of Walter Hayes, Aston Martin started development

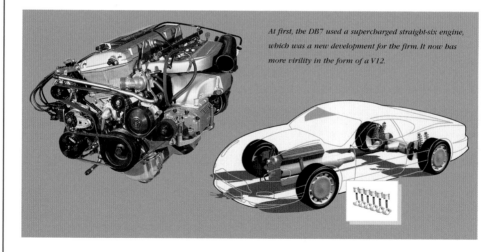

At first, the DB7 used a supercharged straight-six engine, which was a new development for the firm. It now has more virility in the form of a V12.

work on what was dubbed DP1999 but would eventually become the DB7. Ford's acquisition of Jaguar two years after Aston Martin proved fortuitous. Jaguar had been dabbling with new ideas, and Aston Martin was able to take them further.

A NEW RENAISSANCE

Aston Martin's new prototype was effectively created out of an XJS chassis, fitted with a revamped XJ40 engine, and using a lithe new body designed by Ian Callum. The look was totally up to date, but at the same time echoed the classic Aston look of the 1960s. This link was underlined when former boss David Brown was invited back to become life president, hence the DB7 title.

After the Jaguar XJ220 project came to an end, the Oxfordshire factory where it was built was turned over to Aston production. Launched in 1994, the DB7 has become Aston Martin's biggest-selling model ever,

and is still in production today, now fitted with a V12 engine. Although now almost a decade old, and overshadowed by the newer flagship Vanquish, it is still an exhilarating and superior British supercar.

Aston Martin DB7

Top speed:	252.5km/h (157mph)
0–96km/h (0–60mph):	6.0 secs
Engine type:	Straight six
Displacement:	3239cc (198ci)
Max power:	250kW (335bhp) @ 5750rpm
Max torque:	542Nm (400lb-ft) @ 3000rpm
Weight:	1754kg (3858lbs)
Economy:	4.87km/l (13.8mpg)
Transmission:	Five-speed manual
Brakes:	Four-wheel discs, vented at front, solid at rear, ABS
Body/chassis:	Steel floorpan with alloy and composite 2+2 coupe body

UNITED KINGDOM

BENTLEY 4.5 LITRE

All vintage Bentleys have an aura of romance to them, but few are held in the same regard as the Le Mans-winning 4.5 Litre models.

This is the supercharged version of the Bentley 4.5 Litre, known as a Blower. The Roots-type supercharger, designed by tuner Amherst Villiers, was bolted to the front of the car and worked as a sort of pump, forcing air into the combustion chamber to give more fuel/air mixture. It provided a useful increase in power, but put a lot of strain on the engine.

Big drum brakes were fitted on all four wheels, although their efficiency was compromised by cable operation. The large lever on the right of the cockpit is the handbrake, which was adjustable using the screw grip at the bottom of the handle.

There is only a four-cylinder engine under the long, side-opening hood, although each cylinder had four valves, high tech for the era.

Given the sophistication of later Bentleys, prewar models were very crude. The 4.5 Litre had a solid beam axle at the front, and a primitive live axle at the rear. Both were supported by leaf springs.

This stone guard covers the twin SU carburettors. The fuel tank at the rear, and the headlamps, had similar protection.

UNITED KINGDOM

47

UNITED KINGDOM

I n 2003, Bentley won the Le Mans 24-hour Race for the sixth time. It was a historic achievement, the first time the famous British marque had taken the chequered flag at the French road circuit for 73 years.

The modern cars – the EXP Speed 8s – were very different machines to those which had raced during Bentley's heyday in the 1920s. Those thunderous machines, referred to dismissively by Ettore Bugatti as 'the fastest lorries in the world', were epitomized

The Bentley's tall construction meant a long stroke engine, and thus a lot of torque. What was radical at the time, though, was the use of four valves per cylinder. Most cars of this era (and for quite a while afterwards) had just two.

by the 4.5-Litre cars, which won Le Mans in 1928 and supported winning Bentleys in other years.

Powerful but primitive, the 4-.5-Litre was effectively a 6.5-Litre Bentley with two cylinders removed. With its long bonnet, and huge, staring headlamps, it was an imposing vehicle that humbled the opposition both in motion and when standing still. Its primary purpose was racing – and it excelled – but road examples did go to the rich, famous and privileged of the time.

BLOWN AWAY

In 1929, against WO Bentley's wishes, supercharged versions started to be built on a small scale, thanks to the efforts of affluent racer Tim Birkin. A total of 54 Blowers would subsequently be made. These proved less than successful in competition, but their crude brawn and the Bulldog spirit of their drivers turned them into the most legendary of all vintage Bentleys.

By 1931, Bentley lost its independence, taken over by Rolls-Royce, a company more concerned with style than sporting muscle. The 4.5-Litres and Blowers disappeared too.

Bentley 4.5 Litre Supercharged

Top speed:	201km/h (125mph)
0–96km/h (0–60mph):	Not quoted
Engine type:	In-line four
Displacement:	4398cc (268ci)
Max power:	139.5kW (175bhp) @ 3500rpm
Max torque:	179Nm (240lb-ft) @ 2400rpm
Weight:	1925kg (4235lbs)
Economy:	1.77km/l (5mpg)
Transmission:	Four-speed manual
Brakes:	Four-wheel drums
Body/chassis:	Steel ladder frame with cross bracing and open steel and fabric body

UNITED KINGDOM

BENTLEY TURBO R/T

Built for one year only, the Turbo R/T was the ultimate expression of Bentley's turbocharged range. It was not only luxurious, but extremely fast too.

The R/T used Rolls-Royce's enduring alloy V8 engine – as deployed in all post-1959 Rolls-Royces and Bentleys until very recent years – albeit in modified form with Zytec electronic fuel injection and a Garrett turbocharger to achieve more power.

A four-speed automatic transmission was fitted, but it was just as clever as the suspension. It featured adaptive technology, whereby it learned driving styles, and then behaved accordingly when changing ratios.

Top speed of the R/T had to be limited because tyre technology lagged behind the sort of velocities the Bentley could achieve.

In order to give the Bentley a smooth but sporting ride, electronic shock absorbers were fitted. Road effects were monitored by the control system, which then adjusted each shock individually to cope with the stresses being imposed on the suspension. Apart from this, the RT followed standard marque practice of double wishbones at the front and trailing arms at the rear.

The Turbo R/T was distinguishable from other Bentleys by its meshed front grille, a far cry from the palatial chrome radiator surrounds used on the Rolls-Royces versions of these cars.

Prior to its takeover by Rolls-Royce, Bentley was one of Britain's premier sporting marques, with a fertile history of racing success. But, under Rolls-Royce, all that changed. Bentley's image was diluted, and the cars became little more than re-badged Rolls-Royces without all the shiny bits. At one point, RR even gave serious thought to axing one of the greatest names of British motoring.

Thankfully, that did not happen. Instead, Rolls realized how valuable the Bentley

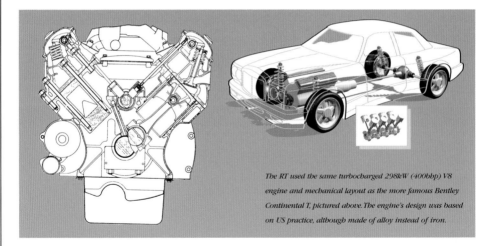

The RT used the same turbocharged 298kW (400bhp) V8 engine and mechanical layout as the more famous Bentley Continental T, pictured above. The engine's design was based on US practice, although made of alloy instead of iron.

badge was, and started building more performance-orientated versions of Rolls-Royces sedans, bearing the Flying B logo.

In 1982, the Bentley Mulsanne (based on the Silver Spirit) was turbocharged with a single Garrett turbocharger, the first proper step on the road to Bentley's revival. In a reference to its normal policy of not disclosing power figures, Rolls-Royce simply stated it was 'sufficient, plus 50 per cent'.

Even more exciting, though, was the Bentley Turbo R of 1985, the R prefix standing for 'road-holding'. There were significant chassis changes to make the handling even more responsive.

THE BEST FOR LAST

It was a bestseller for Bentley, but all good things come to an end. In 1997, with production of the Silver Spirit bodyshell about to finish, Bentley unleashed the most potent Turbo of all. The R/T was an amalgam of the Turbo R and Continental T, featuring the bodyshell and long-wheelbase chassis of the former, and the 4880cc (298ci) torque engine of the latter.

Available for just one year and fabulously expensive, it was a glorious climax to the range that had begun Bentley's renaissance.

Bentley Turbo R/T

Top speed:	244.5km/h (152mph)
0–96km/h (0–60mph):	6.7 secs
Engine type:	V8
Displacement:	6750cc (412ci)
Max power:	298kW (400bhp) @ 4000rpm
Max torque:	664Nm (490lb-ft) @ 2000rpm
Weight:	2477kg (5450lbs)
Economy:	4.39km/l (12.4mpg)
Transmission:	Four-speed manual
Brakes:	Four-wheel discs, vented at front, solid at rear, ABS standard
Body/chassis:	Monocoque four-door saloon

UNITED KINGDOM

BITTER CD

Based on a concept car, the Bitter CD had the exotic looks of a Ferrari or Maserati, but the underpinnings of a Chevrolet...and Opel!

The aerodynamic appearance was aided by the fitting of a large glass hatchback, which also made the car extremely practical compared to most GT sports cars.

Bodywork was steel, constructed by the Stuttgart-based coachbuilder Bauer. Although the CD was a monocoque, it certainly was not a mass-produced car, with each hand-built example taking three weeks to get ready.

Underneath was a shortened Opel Diplomat floor span, with the V8 and all the suspension, power steering and automatic transmission parts carried over. The engine had an American heritage, having originally come from the Chevrolet Corvette, for both Opel and Chevy were General Motors companies. It is a very tuneable unit, and many CDs ended up pumping out more power than they had left the factory with!

The interior was well-appointed, with comprehensive equipment and a wood-panelled dash.

As it was based on an Opel sedan, the CD had space for four inside, although it was less capacious than its less sporting relative thanks to the shortened floor span and sloping rear.

Pop-up quad headlamps were fitted, with auxiliary driving lamps installed underneath, either side of the grille.

oncept cars do not usually make it into production, but are intended purely to be as eye-catching and headline-grabbing as possible. Most are generally impractical for production. However, at 1969's Frankfurt Motor Show, Opel displayed a conceptual coupe that did eventually make the transition to the real world. It would just take a few years, and involve a change of name...

Ex-racing driver Erich Bitter was at that show, saw the Opel, and liked it. A lot. And as

It may have been German in conception and construction, but Bitter's connection with Opel (and thus General Motors) meant that the CD used a Chevrolet Corvette small-block engine (although one not used in the US since 1969).

the German importer of Italian Intermeccanica sports cars – which had associations with Opel – he was in a position to do something about it. By 1971, he had convinced the European GM offshoot to let him build the car under his own Bitter marque name.

Because it had started life as a show car, it required some modifications to make it suitable for road use. Coachbuilder Frua did some of the work, and Erich Bitter a bit more (with the result that there was a certain similarity to the Intermeccanica Indra).

FROM SHOW TO ROAD

The efforts appeared at 1973's Frankfurt Motor Show as the Bitter CD. In flavour it was a luxury grand tourer, in styling it was a supercar. The ravishing shape distracted attention from the fact that underneath were just Opel Diplomat mechanicals, albeit with a Chevy V8 engine. Nevertheless, its handling was good, even if its top speed could not hope to match those it was seeking to rival.

The CD was able to ride out the lean years of the 1970s, although only 395 had been built by 1979, when the SC replaced it.

Bitter CD

Top speed:	209km/h (130mph)
0–96km/h (0–60mph):	9.2 secs
Engine type:	V8
Displacement:	5345cc (327ci)
Max power:	171.5kW (230bhp) @ 4700rpm
Max torque:	427Nm (315lb-ft) @ 3000rpm
Weight:	1750kg (3850lbs)
Economy:	5.66km/l (16mpg)
Transmission:	GM three-speed automatic
Brakes:	Four-wheel discs
Body/chassis:	Integral chassis with steel two-door coupe body

GERMANY

57

BIZZARRINI GT STRADA

Conceived as an extreme racing version of the Iso Grifo, the GT Strada also hit the roads after an argument between Bizzarrini and Iso.

Despite its large rear window and profile, the Bizzarrini was no hatchback. It was fitted with a conventional boot, which was rather on the spatially challenged side.

The first cars had aluminium bodywork, but this was soon changed for fibreglass to make the cars even more lightweight. However, if you really wanted your Bizzarrini to beat the best, you could choose very thin gauge fibreglass panels, with enlarged wheel arches, to make it into a featherweight.

There were distinct similarities in appearance between the Grifo and the GT Strada, especially around the rear haunches. The low-slung Strada – just 117cm (44in) high – looked somewhat like a squashed version of the Iso. Bizzarrini drew up the original design, and then had it refined by styling master Giorgetto Giugiaro of Bertone.

Like the Grifo, the GT Strada also boasted Chevrolet V8 power, using a Corvette 327 unit. Most cars had a single Holley carburettor, but a quad Weber setup was an option, offering much more punch.

The headlamps covered by plastic cowls were an attractive touch that also aided aerodynamics.

ITALY

59

It was Enzo Ferrari's legendary arrogance that resulted in the formation of Lamborghini. Rather less famously, an earlier Ferrari argument led, via a more indirect route, to the formation of another rival, Bizzarrini.

Giotto Bizzarrini had designed the ultra-desirable Ferrari 250 GTO, but a row with Enzo led to him walking out. He then went to Lamborghini, where he designed its V12 engine, and finally he ended up working on a freelance basis for Iso.

Like the Iso on which it was based, the GT Strada used a cast-iron Chevrolet Corvette 327 V8. This was despite the fact that Bizzarrini was a noted engine designer himself.

While the Iso Grifo was being developed, Bizzarrini was working on a racing version. He wanted to build a mid-engined car, but Iso finances dictated against this. Instead, Bizzarrini had to settle on the same front-engined V8 format as the road Grifo, albeit with a shortened chassis and the engine moved as far back as possible (so much so that there was a panel on the dashboard to get to the distributor!).

A WILDER ISO

Both the road Grifo and Bizzarrini's low-slung racing A3/C version were unveiled at the 1963 Turin Motor Show. Then, in 1966, came another argument, this time between Iso boss Renzo Rivolta and Bizzarrini. It turned out that Bizzarrini had registered the Grifo name. For relinquishing it back to Iso, he was given 50 A3/C kits in 'payment'.

And so the GT Strada was born, a road-going version of the A3/C with a touch more comfort inside, and a Chevy 327 V8 engine.

Despite being remarkably fast, and even more dramatic looking than the Iso on which it was based, it failed to make much of an impact on the supercar scene. Just 139 were built up to 1969.

Bizzarrini GT Strada

Top speed:	266km/h (165mph)
0–96km/h (0–60mph):	6.4 secs
Engine type:	V8
Displacement:	5359cc (327ci)
Max power:	272kW (365bhp) @ 6500rpm
Max torque:	466Nm (344lb-ft) @ 4000rpm
Weight:	1150kg (2530lbs)
Economy:	4.95km/l (14mpg)
Transmission:	Four-speed manual
Brakes:	Four-wheel discs
Body/chassis:	Separate pressed-steel chassis with two-door coupe body

ITALY

61

BRISTOL BEAUFIGHTER

Bristol is one of Britain's most eccentric prestige car manufacturers. The Beaufighter was one of its more idiosyncratic offerings, a blunt-edged supercar of questionable appearance.

UNITED KINGDOM

It might come as a surprise to find that the Beaufighter was actually designed by a notable coachbuilder. The Italian styling house of Zagato was behind the square-cut styling of a car that looked like it had been chiselled from a block of wood. One of its less imaginative efforts...

Bristol chose American brawn to power its cars. Under the expansive bonnet was a big Chrysler V8, with a Rotomaster turbo fitted. The transmission was also courtesy of Chrysler, with the well-respected TorqueFlite three-speed system.

Beaufighters – and other Bristols – featured hinged side panels on either side of the front fenders. On the left-hand side was the spare-wheel, while the right one hid the battery, water bottle and other ancillaries.

The Beaufighter had an aluminium body construction, a legacy of Bristol's previous aviation links. It was named after a long-range fighter built during World War II.

It is 'almost' a convertible! The soft rear section of the roof could be folded back, the sunroof removed and the rear quarter windows opened for fresh air motoring. Or the standard-fit air conditioning could just be used!

That Bristol has managed to survive into the 21st century is something of a miracle. While many of its low-volume, high-price contemporaries have fallen by the wayside, Bristol has weathered all the usual ups and downs of the specialist car manufacturer. It has always occupied an exclusive niche, building expensive, high-powered cars of an eccentric, but somehow typically British, nature.

The Beaufighter was a typical example of Bristol's individual approach towards car design. Starting life as the 412, it was already an anachronism when launched in 1975.

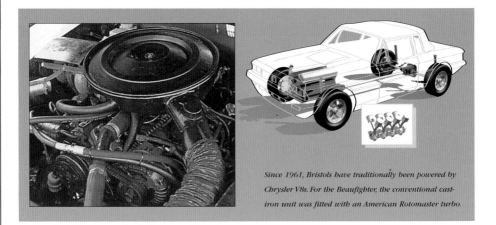

Since 1961, Bristols have traditionally been powered by Chrysler V8s. For the Beaufighter, the conventional cast-iron unit was fitted with an American Rotomaster turbo.

Mounted on a separate chassis (which could trace its origins back to the prewar BMW 327), the body was an uncompromising piece of styling by Zagato. Once known for its luscious curves, the Italian coachbuilder seemed to have taken a house brick as its inspiration for the Bristol project: there was barely a rounded edge to be found. Indeed, the 412 was a piece of angular aggressiveness, shunning accepted beauty for sheer brusqueness.

Those who appreciated Bristol's singular personality loved the alloy-bodied luxury bruiser because it was so different to anything else on the road. With its American V8 power, it was also fast and smooth, not to mention exclusive.

TURBOCHARGED FIGHTER

In 1980, Bristol added a turbocharger to create the fastest accelerating four-seater car in the world. Dubbed the Beaufighter (a tribute to an aircraft from World War II), this even-more mighty Bristol received four square headlamps at the front, as well as a chunkier gearbox.

Production continued until 1992, by which time the Beaufighter looked even odder in comparison to what was around it.

Bristol Beaufighter

Top speed:	241km/h (150mph)
0–96km/h (0–60mph):	5.9 secs
Engine type:	V8
Displacement:	5898cc (360ci)
Max power:	Not disclosed
Max torque:	Not disclosed
Weight:	1750kg (3850lbs)
Economy:	6.3km/l (18mpg)
Transmission:	Torqueflite three-speed automatic
Brakes:	Four-wheel discs
Body/chassis:	Steel chassis with alloy two-door body

BMW 507

One of the most handsome cars ever created, the sublimely proportioned 507 was styled by a German Count, and was the first great post-war BMW.

The entire body was built from lightweight aluminium. Styling was by the German aristocrat Count Albrecht Goertz, who later went on to create the Datsun 240Z.

BMW used its V8 engine for propulsion. Designed by Fritz Fiedler, it featured a robust five-bearing crankshaft, wedge-shaped combustion chambers and light alloy cylinder block and heads. It was used in other BMWs of the era, but for the 507, the 3168cc (193ci) unit had its compression raised to provide more power.

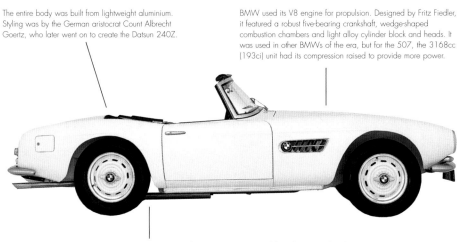

The chassis of the 507 was borrowed from the 502 sedan, but with a hefty 360mm (14in) chopped out of it and some extra strengthening carried out. The suspension also came from the bigger car, although this was revised too.

A stylized version of BMW's ancestral 'split kidney' grille appeared at the front of the 507, running horizontally instead of vertically. A similar look would later appear on the BMW Z8.

In total, there were nine BMW logo roundels decorating the 507.

Servo-assisted drum brakes were fitted at first, but front discs were adopted for later cars.

If you have ever wondered where the inspiration for the current BMW Z8 comes from, look no further than the 507 of 1956 to 1959. With its muscular, curvaceous haunches, chrome and BMW logo-adorned side vents, snarling front grille and general aura of assertiveness, the Z8 was a loving homage to one of the most elegant and beautiful of all BMWs.

The 507 was intended as a car to tackle the all-conquering Mercedes 300SL, as well as boost BMW's sales in America. For that

BMW's V8 engine, designed by Fritz Fiedler, was an advanced design for its era, made out of light-alloy with a five-bearing crankshaft and efficient wedge-shaped combustion chambers.

latter reason, its inspiration came from US sales representative Max Hoffman, who approached Count Albrecht Goertz (then working for an American design agency) to create a two-seater sports car that would find favour in the States.

BEAUTIFUL BEEMER

Initially, there was some resistance to the 507 from high up in BMW, but surely that must have melted when the car was unveiled. It was absolutely gorgeous, one of the prettiest proportioned cars around. And it had substance too. Its high-compression V8 engine could push it to 200km/h (124mph), much faster on some tuned versions, while the close-ratio gearbox ensured sporting acceleration. From 1958, front discs also improved its stopping power.

Unfortunately for BMW, the 507's hugely expensive price meant that sales were always low, with just 253 made up until 1959. This, and its painstaking handbuilt

nature, resulted in the 507 costing the German company money on every one it made. When BMW faced bankruptcy at the end of the 1950s, it was no surprise that the exquisite 507 was an immediate victim of resultant cutbacks.

BMW 507

Top speed:	200km/h (124mph)
0–96km/h (0–60mph):	8.8 secs
Engine type:	V8
Displacement:	3168cc (193ci)
Max power:	112kW (150bhp) @ 5000rpm
Max torque:	236Nm (174lb-ft) @ 4000rpm
Weight:	1290kg (2840lbs)
Economy:	4.89km/l (13.8mpg)
Transmission:	Four-speed manual
Brakes:	Four-wheel drums, later front discs
Body/chassis:	Separate steel chassis with two-door aluminium convertible body.

GERMANY

69

BMW M1

BMW has only ever built one mid-engined supercar. The M1 was originally intended as a racing car, with less than 400 built for road use.

The M1 had identical-looking vents on either side of its engine hatch, but they served different purposes. The ones on the right-hand side drew breathing air into the engine, while those on the left allowed heat to escape.

Behind the M1's angular lines was Giorgetti Giugiaro of Ital Design. The dramatic wedge-shape took its inspiration from a BMW Turbo Coupe concept car that had appeared on the show circuit in 1972.

Lamborghini was originally commissioned to design and build the M1's chassis, but ran into financial problems. Eventually, the Marchesi firm, in nearby Modena, built them.

Surprisingly for a mid-engined supercar, the engine was not a 'V' configuration. Instead, BMW used a version of its straight-six engine dating from the 1960s, modified with a twin-cam head and 24 valves. And instead of alloy, it had a heavier cast-iron block.

BMW's traditional grille appeared in miniature form on the front of the M1, keeping both the oil cooler and radiator fed with fresh air. Retractable headlamps were a first for the marque.

All the body panels were made of fibreglass.

GERMANY

71

GERMANY

Not even BMW gets it right all the time. The development of its M1 supercar was accompanied by a catalogue of problems, delays and regulatory headaches. The car was conceived to contest the Group 5 racing series, but by the time it finally made it into production, almost a decade later, it had been superseded by younger upstarts. Frankly, the M1 was a bit of a disaster...

Still, every cloud has a silver lining, although it was BMW's customers who saw

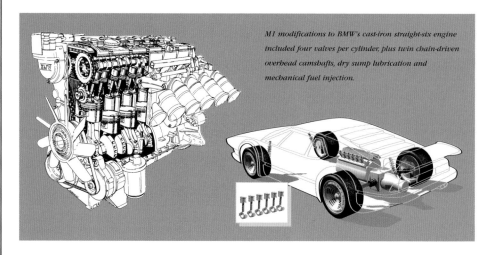

M1 modifications to BMW's cast-iron straight-six engine included four valves per cylinder, plus twin chain-driven overhead camshafts, dry sump lubrication and mechanical fuel injection.

it rather than the firm itself: 397 homologation models of the M1 found their way into private hands, and it became one of the greatest supercars of them all.

GERMAN POWER, ITALIAN DESIGN

In 1972, BMW unveiled a show car which so impressed its Motorsport division that it pressed for the car to be built as a racing machine. Germany turned to Italy to bring its new idea to life. Giorgetti Giugiaro, of Ital Design, was asked to style the fibreglass body, while Lamborghini was approached to design the mid-engine chassis. Work on the engine stayed in Germany, with BMW's regular straight-six engine providing the basis for the M1's motive power. Its iron block was retained, but a new 24-valve head was fitted. Fuel was delivered via mechanical fuel injection.

Problems started when a cash-strapped Lamborghini found itself unable to build the chassis. Then, when the M1 did come out in 1979, BMW suddenly found itself having to build 400 road cars under new rules. This it managed by 1981, and finally took the M1 racing…only to find it completely outclassed!

BMW M1

Top speed:	261km/h (162mph)
0–96km/h (0–60mph):	5.7 secs
Engine type:	Straight-six
Displacement:	3453cc (211ci)
Max power:	206.5kW (277bhp) @ 6500rpm
Max torque:	324Nm (239lb-ft) @ 5000rpm
Weight:	1419kg (3122lbs)
Economy:	4.57km/l (12.9mpg)
Transmission:	ZF five-speed manual
Brakes:	Four-wheel vented discs
Body/chassis:	Fibreglass two-door, two-seat coupe body with tubular steel chassis

GERMANY

BUGATTI TYPE 57

Sporting some of the most artistic and exotic coachwork ever produced, Bugatti's Type 57 has been hailed as the greatest of all prewar sports cars.

A variety of different bodies were fitted on the Type 57 chassis, both by Bugatti and independent coachbuilders. The curvaceous Atlantic body, with its swept-back rear bodywork and central fin was the most enticing, but the Atalante, as seen here, almost matched it for desirability. It was built by the Gangloff company, to a design by Bugatti.

The Type 57 had a straight-eight engine (on the SC version, it was fitted with a Roots-type supercharger) with twin camshafts. To get around the problem of blown head gaskets, the engine block and head were made as one complete unit.

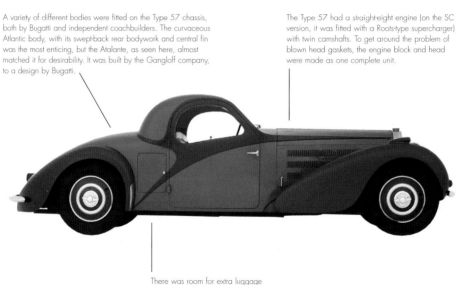

There was room for extra luggage space behind this hatch.

Rear visibility was compromised by the tiny split rear window. Many prewar cars had screens of this type, but on the Atalante, they were even smaller than usual due to the design of the bodywork.

The horseshoe grille was a well-known Bugatti emblem. It had grown out of the earlier egg-shaped devices seen on the first Bugattis. Ettore Bugatti, who had received an art education, described the egg as 'the most perfect shape in nature'.

FRANCE

75

The Type 57 is the best known of all the Bugatti models, thanks in particular to some of the astounding bodywork inventions that ended up decorating its distinguished chassis. Even in the Art Deco environment of the 1930s, when fashions and imagination ran riot, Type 57s looked like nothing else.

Although Ettore Bugatti was Le Patron of the marque, it was his son Jean who created the Type 57 (the number denoting that it was the 57th Bugatti design). Examining all

Fitted with just a single carburettor, the Bugatti straight-eight could produce 104kW (140bhp). However, with a Roots-type supercharger bolted on – as with the 57C and 57SC – it generated 148kW (200bhp).

aspects of the car, he even revamped the company's straight-eight engine, fitting it with twin camshafts. One thing he could not do, though, was alter the suspension. Ettore Bugatti did not believe in independent front suspension, and refused to have it on any of his cars. So the Type 57 perpetuated the old beam axle and leaf springs arrangement.

YEARS OF CHANGE

The original Type 57 was seen in 1934, followed by several different variants. A year later came the 57S, a shortened chassis model with a tuned engine. The 57C soon afterwards was a supercharged version, but the zenith of the series was the 57SC of 1937. This combined the short chassis model with the supercharged engine to create a car capable of 193km/h (120mph). Mechanically, the most significant change to affect all models was the adoption of hydraulic brakes and telescopic shocks on the Series 3 of 1938.

Manufacture was halted by the war, and never restarted. It was a car just too evocative of the glamorous 1930s, and would have seemed out of place in the austere post-war world.

Bugatti Type 57 SC

Top speed:	193km/h (120mph)
0–96km/h (0–60mph):	10.0 secs
Engine type:	In-line eight
Displacement:	3257cc (199ci)
Max power:	104kW (140bhp) @ 4800rpm
Max torque:	Not quoted
Weight:	967kg (2127lbs)
Economy:	3.89km/l (11mpg)
Transmission:	Four-speed manual
Brakes:	Four-wheel drums
Body/chassis:	Ladder-type steel chassis with two-door coupe bodywork

FRANCE

BUGATTI EB110

It was a brave, but ultimately doomed, decision to restart Bugatti 40 years after Ettore Bugatti had died, although it did result in the scintillating EB110.

A close-ratio six-speed transmission was necessary to handle all the power and torque generated by the V12, fitted just in front of the mid-mounted, longitudinal engine.

Marcello Gandini – most associated with Lamborghini – was hired to style the EB110, but after arguments with Bugatti's management about its design, he left and asked for his name to be removed from the car. It was finished by an Italian architect!

The EB110 was so wide that the huge doors opened upwards.

Most EB110s were blue – the traditional
Bugatti colour – although a limited number
appeared in silver.

By supercar standards, the size of the EB110's V12 engine was
small, at just 3500cc (213.5ci), but its power outputs – between
412kW and 455.5kW (553–611bhp) – were enormous, thanks
to its high-revving ability and four turbochargers of Japanese origin.

You have to look closely, but Bugatti's
trademark horseshoe radiator grille was
continued on the front of the EB110, albeit
in miniature form

Permanent four-wheel drive was used
on the EB110, although there was a
rear-wheel drive bias in the power
distribution.

ITALY

E ttore Bugatti, the enigmatic founder of the marque that bore his name, died in 1947. By 1956, Bugatti the company had gone as well.

Then came a wealthy Italian businessman, who decided that the time was right for a revival and bought the name. Fired by lofty ideals, he built a new factory in Italy, hired some of the best men in the industry to work for Bugatti, and, most crucial of all, announced the production of a new V12-engined supercar, the EB110.

Although the capacity of the V12 was small, it put out an excessive amount of power, thanks to the use of four Japanese IHI turbochargers. Fitting four also overcame turbo lag.

A TRIBUTE TO ETTORE

The letters stood for Ettore Bugatti's initials, the 110 denoting the fact that the car was due to be launched on what would have been his 110th birthday in 1991. Marcello Gandini, the man responsible for the Lamborghini Miura, Countach and Diablo, began work on styling, but after constant meddling with his ideas, he told Bugatti, 'If you want more changes, do them yourself!'

Bugatti did, and what finally appeared was controversial, but undeniably daring, with its close set headlamps nestling under shrouds in the short, sloping nose. Technologically, it was a tour de force, having a mid-mounted V12 engine of surprisingly small capacity, but boosted by four turbochargers to deliver an incredible amount of power. If that were not enough, there was also the EB110S, with even more potency, intended for GT racing.

But the 'new' Bugatti was in trouble. Sales failed to reach expectations - even though racing driver Michael Schumacher bought one - and the line came to an end again in 1995. The marque is now owned by the Volkswagen Audi Group.

Bugatti EB110

Top speed:	341km/h (212mph)
0–96km/h (0–60mph):	3.5 secs
Engine type:	V12
Displacement:	3500cc (213.5ci)
Max power:	411.5kW (552bhp) @ 8000rpm
Max torque:	732Nm (450lb-ft) @ 3750rpm
Weight:	1623kg (3571lbs)
Economy:	6.37km/l (18mpg)
Transmission:	Six-speed manual
Brakes:	Four-wheel vented discs
Body/chassis:	Alloy two-body, two-seat coupe with carbon fibre monocoque chassis

ITALY

81

CATERHAM SUPER 7

It was the great Colin Chapman of Lotus who originally designed the Seven, but Caterham which took it over and turned it into a savage supercar.

Underneath the minimalist bodywork is essentially the same steel tube backbone spaceframe with which the Seven started off in the 1950s. Caterham has strengthened it over the years, though. The rollover bar forms part of the chassis tubing.

Bodywork is a mixture of aluminium and fibreglass panels (the nose cones and fenders).

According to Colin Chapman, he designed the body of the original Lotus Seven in a week. That was back in 1957, and it has been in production ever since. Not a bad run for seven day's work!

Just visible behind the right-hand front fender are the twin Weber carburettor air cleaners. They had to stick out of the hood, because space in the Seven's long but narrow engine bay was so limited.

In order to mate the Ford Sierra five-speed
transmission with the Vauxhall engine, Caterham
had to design a special bellhousing.

The Seven has always used parts from whatever other
cars were available at the time, with the engines
traditionally coming from Ford or Rover. On this 1990s
Super Seven HPC, a Vauxhall (Opel) Astra four-cylinder
twin-cam 1998cc (122ci) did service.

It is the unlikeliest of all supercars. Born in the late 1950s as a low-budget racer, Graham Chapman's Seven was responsible for establishing Lotus as a respected sports manufacturer, becoming a motoring icon and British institution in the process.

When Lotus moved onto other things, the rights to the design were bought by Graham Nearn of Caterham Cars in 1973. His firm had sold the Lotus version, now he wanted to build his own Super Seven.

Wisely, Caterham chose not to mess with a successful formula. The Seven had always

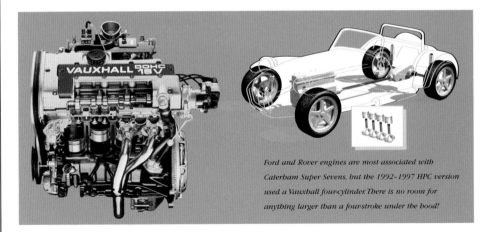

Ford and Rover engines are most associated with Caterham Super Sevens, but the 1992-1997 HPC version used a Vauxhall four-cylinder. There is no room for anything larger than a four-stroke under the hood!

been about driver enjoyment, and nothing was allowed to spoil that. Spartan in the extreme, with primitive bodywork and no weather protection, the raison d'être of the Seven was just to go fast, whether on straights or bends. As basic as it was, the car was a massive bundle of four-wheeled fun.

FASTER THAN FERRARI

What Caterham did do was slowly improve the Seven, using modern components to enhance the classic original, or simply making its own. As well as Ford engines, the company branched out into Vauxhall and Rover K-series power sources. Some of the hairier versions Caterham would eventually come up with (such as the JPE) could out-accelerate Ferraris…

That kind of performance put them solidly into supercar territory, and although the average Ferrari or Aston Martin owner would not look twice at a Super Seven, for others, the raw Caterham offered unparalleled

performance at the kind of price even mere mortals could manage…just. The Caterham Super 7 is still built today, and probably will be for decades to come.

Caterham Super 7 HPC

Top speed:	203km/h (126mph)
0–96km/h (0–60mph):	5.4 secs
Engine type:	In-line four
Displacement:	1988cc (122ci)
Max power:	130.5kW (175bhp) @ 6000rpm
Max torque:	210Nm (155lb-ft) @ 4800rpm
Weight:	629.5kg (1385lbs)
Economy:	6.90km/l (19.5mpg)
Transmission:	Ford five-speed manual
Brakes:	Four-wheel discs
Body/chassis:	Tubular steel spaceframe with aluminium honeycomb, alloy sheet and fibreglass bodywork

UNITED KINGDOM

85

CHEVROLET CORVETTE

As America's first proper sports car, the Corvette has become a legend. With fibreglass bodies and fuel-injected V8 engines, the 1955–62 models had supercar performance.

The Corvette was the first mass-produced US car to have a fibreglass body. The process was still unrefined, and the Corvette had a few rough edges, but the lovely shape made up for those. Fibreglass was adopted because it made it easier to get the first cars out to customers quickly when the decision was taken to put the Corvette into production. Yet the material is still used on the current version.

The first generation of Corvettes – 1953 to 1955 – were built with slab sides, but the most distinctive feature of the 1956 restyle was these scallops hewn out of the sides. Having them painted a contrasting colour to the rest of the car was an $18 option, but one many customers took because the car looked so good as a result.

Small air scoops were fitted on the top of each front fender to feed additional air into the engine bay.

Chevrolet's V8 engine was fed by carburettors until 1957, when fuel-injection was fitted, which provided additional power. As with the fibreglass bodies, Chevrolet's use of injection was pioneering.

UNITED STATES

87

UNITED STATES

I ronically, if it had not been for Ford's Thunderbird, Chevrolet's American icon, the Corvette, might have died out after a few brief years. It started out as a Harvey Earl show car in 1952, and so captured the imagination of the public that Chevrolet rushed it into production for 1953. But the initial enthusiasm for this fibreglass sports car turned lukewarm when its straight-six engine proved to have disappointing performance.

Then Ford launched its V8-powered two-seater Thunderbird in 1955. It was a challenge Chevrolet could not refuse. It

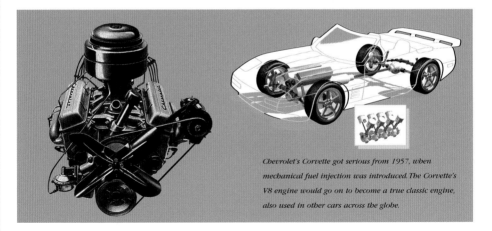

Chevrolet's Corvette got serious from 1957, when mechanical fuel injection was introduced. The Corvette's V8 engine would go on to become a true classic engine, also used in other cars across the globe.

tweaked the Corvette's styling – much for the better – and dropped in its own V8. Suddenly, its plastic fantastic sports car was capable of more than just taking on the Thunderbird. It could even give most European supercars a run for their money.

BEATING THE 'BIRD

When the Thunderbird grew bigger and less sporting for 1958, the Chevrolet stayed keen, lean and mean. In fact, it got even better. The inspiration of Chevy engineer Zora Arjus Duntov resulted in fuel injection being added to the 'Vette. Power and performance shot up, and sales followed suit.

Chevrolet made improvements to its two seater every year. Some were subtle, others were more noticeable, such as the quad headlamp arrangement introduced in 1958. In 1963, the delicious curves were replaced by the angular lines of the Corvette Sting Ray, designed by Harvey Earl's successor at Corvette, Bill Mitchell.

The Corvette has been made ever since, under various styling guises. But it is the 1955–62 series that are among the best loved, not just by American enthusiasts, but sports car fans worldwide.

Chevrolet Corvette

Top speed:	212km/h (132mph)
0–96km/h (0–60mph):	5.7 secs
Engine type:	V8
Displacement:	4342cc (283ci)
Max power:	211kW (283bhp) @ 6200rpm
Max torque:	366Nm (270lb-ft) @ 3600rpm
Weight:	1241kg (2730lbs)
Economy:	4.60km/l (13mpg)
Transmission:	Close ratio three-speed manual
Brakes:	Four-wheel drums
Body/chassis:	Welded steel box section chassis with fibreglass two-door roadster body

UNITED STATES

CITROËN SM

The SM was a gloriously eccentric car, the result of a collaboration between Citroën and Maserati to build a supercar brimming with performance, luxury and character.

Its swept-back rear bodywork left no room for a conventional boot, so the SM was a hatchback, an uncommon feature for a supercar.

Despite the Maserati influence, the look of the SM was pure big Citroën, echoing the DS, and providing clues to the forthcoming CX and GS models. Its designer was Robert Opron.

Original plans called for the V8 engine to power the SM, but it did not fit, and was also too powerful! So, two cylinders were chopped off the end to turn it into a V6. The crankshaft had to be heavily counterweighted by Maserati engineer Giulio Alfieri to make up for the missing two cylinders. The gearbox was mounted in front of the engine.

Set between the two cylinder heads of the V6 engine was the hydraulic pump, a vital piece of equipment, providing pressure for the hydropneumatic self-levelling suspension, Vari-Power self-centring power steering and the ultra-sensitive brakes.

The entire nose of the car, including the number plate, was enclosed by glass. The inner two headlamps were linked to the steering and turned when the car went around corners.

FRANCE

Citroën has always had a reputation for being...well, unconventional. In fact, if there was an established way of doing things, Citroën would do it differently, simply to prove it could.

So, when it took over the Italian supercar manufacturer Maserati in 1968, it seemed to be offering an intriguing promise that something very special was on the way. Enthusiasts who relished Citroën's unorthodox approach had to wait only until 1970 to find out what.

That year, a surprised world was treated to the SM (Serie Maserati). In terms of styling

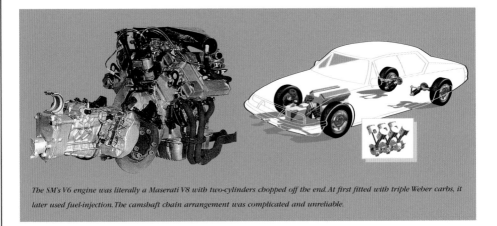

The SM's V6 engine was literally a Maserati V8 with two-cylinders chopped off the end. At first fitted with triple Weber carbs, it later used fuel-injection. The camshaft chain arrangement was complicated and unreliable.

and downright quirkiness, it was everything everybody expected of Citroën. Inside, outside and under the skin, the elegant and angular grand tourer was a masterpiece of technical ingenuity. A Maserati-designed V6 engine put power through the front wheels, while a complicated system of hydraulics controlled the steering, suspension and brakes, and even swivelled the headlamps.

ALL TOO MUCH

In fact, it was all too complicated. The rush to get the SM into production resulted in shortcomings in the design and reliability problems. The press and the French adored the car, but everybody else seemed suspicious of its novelty. Sales, strong at first, faltered, particularly in the wake of the 1973 fuel crisis. Even the introduction of fuel-injection did not help matters.

When Citroën was taken over by the more conventional Peugeot in 1974, the writing was on the wall for the SM. Legend has it that when the new owner stopped production of the SM in 1975, the order also went out for hundreds of uncompleted cars to be destroyed as well.

Citroën SM

Top speed:	228.5km/h (142mph)
0–96km/h (0–60mph):	8.5 secs
Engine type:	V6
Displacement:	2670cc (168.5ci)
Max power:	133kW (178bhp) @ 5500rpm
Max torque:	232Nm (171lb-ft) @ 4000rpm
Weight:	1453kg (3197lbs)
Economy:	4.39km/l (12.4mpg)
Transmission:	Five-speed manual
Brakes:	Four-wheel discs
Body/chassis:	Unitary construction with two-door, four-seat hatchback body

FRANCE

93

CORD 812 SUPERCHARGED

Looking more like an Art Deco ornament than a car, the supercharged Cord 812 was an astonishing creation that was not just revolutionary on the surface.

The supercharged 812s could be distinguished from their more lowly stablemates by the flexible exhaust piping exiting from underneath the hood on each side. There was no real mechanical or performance bonus in doing this, but it did look very sexy!

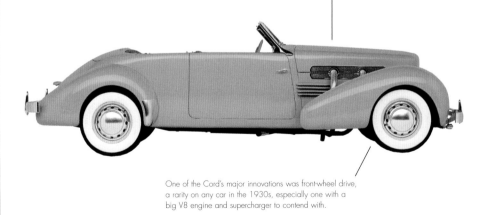

One of the Cord's major innovations was front-wheel drive, a rarity on any car in the 1930s, especially one with a big V8 engine and supercharger to contend with.

A manual Bendix 'Electric Hand' preselector gear box (where the gears are selected before the clutch is depressed) was fitted, operated by electro-vacuum due to its distance away from the driver, underneath the engine. A small lever beside the steering wheel was all that was needed to go through the four forward ratios.

Continuing the pioneering theme, the headlamps (which were actually aircraft landing lights) were retractable, operated by a wind-up handle on the dashboard. For reasons best known to Cord, it placed the crank over on the far side of the passenger side, meaning the driver either had to stop to put the headlamps up, or risk an accident!

UNITED STATES

The 1930s was a flamboyant decade for American design, with Art Deco influencing even the most humble of objects. The automotive industry wholeheartedly embraced the new look, and the era saw many extravagant vehicles take to the streets. Form was definitely more important than function.

But few could compare to the supercharged Cord 812S for sheer excessiveness. It was years ahead of its time, both mechanically and aesthetically. What made it even more extraordinary was that it was a last gasp effort by a dying company.

Originally planned as a Duesenberg (another branch of the Cord Corporation),

Lycoming, an aircraft manufacturer that was part of the same family as Cord, built the V8 engines for the 812, using side valves in L-heads operated by a single camshaft.

the expensive Cord 810 was launched in 1935. It was a very special car, packed with innovation. There was front-wheel drive, a four-speed electro-vacuum transmission, independent front suspension, hydraulic brakes, pop-up headlamps and an aviation-inspired dashboard. On their own, any of these features would have grabbed headlines, but what really aroused interest was its remarkable and unique coffin-nosed styling by Gordon Beuhrig.

SUPERCHARGING ARRIVES

Too much was new, though, and putting right problems for customers cost Cord dear. In 1937, the car was relaunched as the 812. A Schwitzer-Cummins centrifugal supercharger was added to the V8 engine, offering the performance that its appearance – now with chromed exhaust pipes sprouting from each fender – deserved.

This did not help, however. Just over 1100 models were built before Cord's financial problems overwhelmed it, and production of the 812 drew to a close late in August 1937.

Cord 812 Supercharged

Top speed:	178.5km/h (111mph)
0–96km/h (0–60mph):	13.8 secs
Engine type:	V8
Displacement:	4729cc (288ci)
Max power:	141.5kW (190bhp) @ 4200rpm
Max torque:	368.5Nm (272lb-ft) @ 3000rpm
Weight:	1868kg (4110lbs)
Economy:	6.37km/l (10mpg)
Transmission:	Four-speed manual
Brakes:	Four-wheel hydraulically-operated drums
Body/chassis:	Welded steel floorpan and side rails with two-door coupe or convertible body

UNITED STATES

DE TOMASO MANGUSTA

Under no circumstances could the Mangusta be called a civilized car. It was a brutally fast concoction of American V8 power married with Italian engineering.

De Tomaso used bought-in Ford HiPo 289 and 302 V8 engines to power the Mangusta. The HiPo 289 was also used in the Shelby GT350, with De Tomaso having the same modifications done to its unit as Shelby had to its own. The ZF five-speed transmission was mounted behind the engine.

Giorgetto Giugiaro came up with the design of the Mangusta during his spell as head of styling at Ghia, which was then also owned by De Tomaso. It was one of his more intimidating efforts, the look of the car perfectly reflecting its aggressive temperament.

The rear hood owed a lot to vintage car design. Instead of one lid which opened as a whole, each half could be opened, hinged from the centre. This did little for ease of access, though.

Because the mid-mounted V8 engine was made of iron (most supercars used alloy), and the ZF transmission was also at the rear, the Mangusta was very unbalanced, with a 68:32 bias towards the back of the car. This badly affected handling. Using larger rear tyres helped matters only slightly.

Mangusta means 'mongoose', and it is said that the only animal capable of beating a cobra is the mongoose. Alejandro de Tomaso pitched his Mangusta directly against the AC Cobra, but despite using the same Ford V8 as its adversary (albeit mid-mounted), his Italian supercar failed to vanquish its Anglo–American opponent.

On the surface, the Mangusta had a lot going for it. It was a brawny-looking car, dripping with attitude thanks to a no-nonsense design by Giugiaro. Its Ford V8 engine, with 228kW (306bhp), promised a

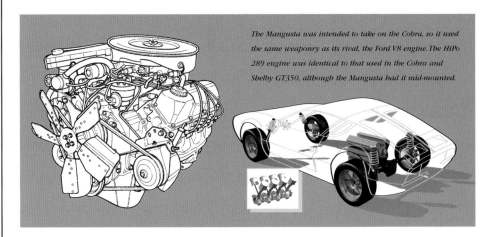

The Mangusta was intended to take on the Cobra, so it used the same weaponry as its rival, the Ford V8 engine. The HiPo 289 engine was identical to that used in the Cobra and Shelby GT350, although the Mangusta had it mid-mounted.

great deal. But while the Mangusta was incredible in a straight line, it was less good when it came to corners.

THE MONGOOSE'S TAIL

Its chassis, carried over from the previous Vallelunga model, was prone to flexing and shaking. And with its major mechanical components installed at the rear, and only a lightweight alloy boot lid and radiator to anchor down the front, handling was erratic at best – and downright dangerous at worst!

The Mangusta debuted at the Turin Motor Show of 1966, with its body made out of fibreglass. By the time it went into production in 1967, alloy and steel had replaced the plastic, but apart from that, it was largely unchanged.

However, its production life was short, with just 401 being made before it was replaced in 1971. Most of those ended up in the United States, where their V8 engines were strangled compared to the European spec versions. The Pantera superseded the Mangusta, lasting far, far longer and selling much, much more.

De Tomaso Mangusta

Top speed:	249.5km/h (155mph)
0–96km/h (0–60mph):	7.0 secs
Engine type:	V8
Displacement:	4950cc (302ci)
Max power:	228kW (308bhp) @ 4800rpm
Max torque:	420Nm (310lb-ft) @ 2800rpm
Weight:	1386kg (3050lbs)
Economy:	4.42km/l (12.5mpg)
Transmission:	Rear-mounted five-speed ZF manual
Brakes:	Four-wheel Girling discs
Body/chassis:	Sheet steel backbone chassis with engine and transmission as stressed members, and alloy and steel two-door coupe body

ITALY

DODGE CHARGER DAYTONA

It is impossible to mistake the Dodge Charger Daytona for anything else...unless it was a Plymouth Superbird! Welcome to one of America's most eye-catching cars.

The aerodynamic 'droop snoot' nose cone was made out of fibreglass and extended the length of the standard Charger by 432mm (17in). Because it covered up the usual headlamps set alongside the grille, quad pop-up lamps were installed.

Daytonas were based on standard Chargers, and were created to make them more competitive at NASCAR racing. The name is a tribute to one of the most famous of American racing circuits.

Apart from the nose cone, the rest of the Daytona was of monocoque steel construction. The front extension never seemed to fit very well.

Spoilers did not get more outrageous than that of the Daytona. Not that this was all show. The wing was 610mm (24in) tall in order to allow the boot to be opened properly, and air flow was less impeded over it. The surface was adjustable to vary effectiveness.

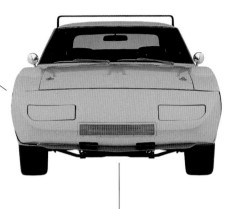

Two V8 engine options were offered, a 440 or the legendary 426 V8 Hemi. Road cars were less finely tuned than their racing counterparts and could not quite manage the 322kp/h (200mph) speeds of their competition counterparts.

UNITED STATES

UNITED STATES

In 1969, man first landed on the Moon. And back on earth, one American auto firm was building the four-wheeled equivalent of the Apollo moon rocket, to try and dominate NASCAR racing. Or at least to end a run of humiliating defeats.

The Dodge Charger Daytona may not have been one giant leap for mankind, but it was a bold step for the Chrysler Corporation. One of the most extremely styled cars ever to be sold to the public, it was pure motoring madness, not so much a vehicle as a macho

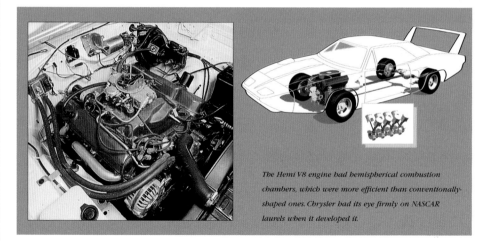

The Hemi V8 engine had hemispherical combustion chambers, which were more efficient than conventionally-shaped ones. Chrysler had its eye firmly on NASCAR laurels when it developed it.

attitude with a steering wheel attached. When describing this very special Charger, it was all too easy to run out of superlatives.

Daytonas were built with one reason: to beat Ford at NASCAR. With its Torino Talladega and Mercury Cyclone Spoiler, the Blue Oval was wiping the asphalt with Chrysler. Understandably, the corporation wanted to get its own back.

DAYTONA DESTINY

Revenge came in the awesome shape of the Dodge Charger Daytona. Big, bad and brash, the Daytona was a Charger 500 with aerodynamic aids tacked on to make it slip through the air faster than a Ford. It worked. Out on the NASCAR circuits, one Daytona hit 323km/h (201mph).

Five hundred street examples had to be built for homologation purposes, and although they were less volatile than their track counterparts, they were the car to own if you craved attention. With garish colours such as Lemon Twist, Plum Crazy and Go Mango, the Daytona was unmissable.

It lasted only a year, before its 'King of Cool' crown was handed to the visually-similar Plymouth Superbird.

Dodge Charger Daytona

Top speed:	217km/h (135mph)
0–96km/h (0–60mph):	5.0 secs
Engine type:	V8
Displacement:	6974cc (426ci)
Max power:	317kW (425bhp) @ 5600rpm
Max torque:	664Nm (490lb-ft) @ 4000rpm
Weight:	1669kg (3671lbs)
Economy:	3.89km/l (11mpg)
Transmission:	Four-speed manual
Brakes:	Four-wheel drums
Body/chassis:	Unitary monocoque construction with steel body panels and fibreglass nose section

UNITED STATES

DODGE VIPER

An unashamed reinvention of the legendary AC Cobra concept, the Dodge Viper is America's best-known supercar, and the only V10-engined road car in production.

All the body panels are of reinforced plastic, mounted on a tubular steel chassis by bonding or bolts.

The Viper is, famously, the world's only car with a V10 engine. At 7998cc (488ci), it is also one of the largest around. During the time the Viper was in development, Dodge's parent company Chrysler also owned Lamborghini, and used the Italian supercar manufacturer to work on the all-aluminium engine.

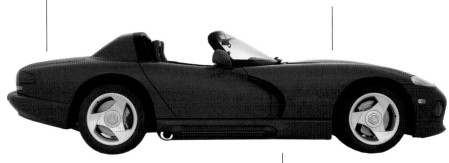

These vents in front of the doors allowed hot air to escape from the engine bay.

Coupe and convertible versions of the Viper are available. On convertibles, the roll-over bar is built into the structure, since it also helps stiffen the body, and the rear window can be easily removed if wanted. A distinctive double curved hard-top, with a dip in the centre, was available for those days when the bad weather just got a bit too much!

Tyres are different sizes. The ones at the rear are 75mm (3in) wider than the front ones, as they had to deal with all the power being transmitted to the road.

The Dodge Viper is one of the most brutal supercars available today. It is an outrageous machine, bristling with aggression and overflowing with power. For sheer posing value, it can take on anything Europe has to offer. It is the sort of car that is not so much driven as tamed.

Originally, the Viper was a concept car. Displayed at the 1989 Detroit Motor Show, it was intended to be a 1990s reincarnation of the AC Cobra. Indeed, the involvement of Carroll Shelby (the Cobra's father) and Lee Iacocca (creator of the Ford Mustang) reinforced the idea that Dodge had decided

It may be derived from a truck engine, but Lamborghini's work on the V10 for the Dodger transformed it into sophisticated monster. It is still the only V10 car engine around, which gives the Viper a lot of its appeal.

to build a modern sports car that had old school values.

Dodge changed its mind about keeping the Viper as a one-off when it saw how enthusiastically it was received. Seeing a way to inject some glamour into the staid image of the Chrysler Corporation, management announced that the car would go into production in 1992.

PLASTIC FANTASTIC

What emerged was no watered-down version of the show car, but a full-blown street racer with big muscles to flex. The V10 engine was taken from a truck, but then worked over by Lamborghini to make it more suitable for a supercar. Indeed, it accelerated like the space shuttle, thanks in part to its plastic bodywork which made it very light.

Originally available just as a convertible, a coupe version – the GTS – joined its topless sister in 1996, with increased power. This,

and its better aerodynamics, made it even faster still.

For 2003, the Viper was revamped, the bodywork modified but just as dramatic.

Dodge Viper

Top speed:	245km/h (152mph)
0–96km/h (0–60mph):	5.4 secs
Engine type:	V8
Displacement:	7998cc (488ci)
Max power:	298kW (400bhp) @ 4600rpm
Max torque:	661Nm (488lb-ft) @ 3600rpm
Weight:	1580kg (3477lbs)
Economy:	12mpg (4.3km/l)
Transmission:	Six-speed manual with electronic shift lockout
Brakes:	Four-wheel vented discs
Body/chassis:	Tubular-steel chassis with two-seat fibreglass reinforced plastic convertible or coupe body

UNITED STATES

DUESENBERG SJ

There have been few cars as imposing as the Duesenberg SJ. It was advertised as 'The world's finest motor car', and its extravagance could humble a Rolls-Royce.

Despite its length, the SJ was a two-seater. The rear boot did contain a rumble seat, but this was hardly the most comfortable of ways to travel.

Pedestrian safety was not an important issue in prewar days, as this nasty-looking spiked radiator mascot demonstrates.

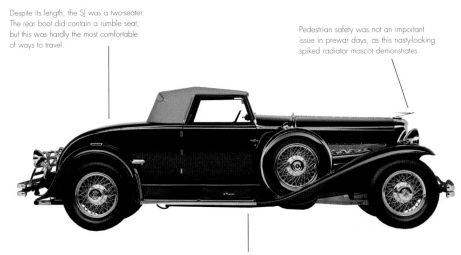

The aluminium dashboard, with its abundance of gauges, looked like it had been borrowed from the cockpit of an aeroplane.

What made the SJ particularly special was its centrifugal supercharger, which boosted power to 239kW (320bhp). Even in the 21st century, this type of output is impressive, but for the 1930s, it was nothing short of a revelation. Nothing could match a SJ for power, and the engines had to be specially modified to handle it.

Duesenberg was part of the Cord Corporation, which also owned the Lycoming aircraft company. It built the in-line eight cylinder engines for the SJ, which were highly advanced for their era. They had twin overhead camshafts, and four valves per cylinder. Another aviation practice was the use of aluminium parts, both inside the engine and for its ancillaries.

UNITED STATES

111

As awe-inspiring as the Duesenberg SJ is, it is a car with a nasty secret. It was an SJ that was indirectly involved in the death of company founder Fred Duesenberg. When the Cord Corporation took over the car firm in 1926, Fred was retained as chief engineer. In 1932, while returning from New York to Indianapolis in a new SJ, he was involved in an accident and later died of his injuries. This is the only blot spoiling the career of what is otherwise a spectacular machine. Calling the Duesenberg

Not only did the Duesenbergs straight-eight engine look stylish, but it was staggeringly powerful and strong as well, thanks to the supercharger and tubular steel con rods.

SJ the greatest American car of the prewar era is no extravagant overstatement. It was an exceptional machine, dazzling to look at, and dazzling to drive, although the ageing chassis was a disappointment. Nevertheless, nothing else on the road at the time was remotely comparable.

THE ROLLS-ROYCE OF AMERICA

The Duesenberg Model J came out in 1928, and was built without worry about cost. Intended to surpass even Rolls-Royce, it did not quite have the class of the British thoroughbred, but did have the luxury, power and sheer flamboyance to compete.

In 1932, the SJ was created by adding a supercharger, which made it the world's most powerful production car. With its chrome-plated exhaust headers emerging from the front fenders, it was an exclusive car designed to attract attention. Actors Gary Cooper and Clark Gable owned specially shortened models.

The Duesenberg SJ had a tragically short life. When the Cord Corporation collapsed in 1937, the Duesenberg went down with it. Just 36 SJs were ever made, each of them very special.

Duesenberg SJ

Top speed:	209km/h (130mph)
0–96km/h (0–60mph):	8.5 secs
Engine type:	In-line eight
Displacement:	6882cc (420ci)
Max power:	239kW (320bhp) @ 4200rpm
Max torque:	576Nm (425lb-ft) @ 2400rpm
Weight:	2273kg (5000lbs)
Economy:	3.54km/l (10mpg)
Transmission:	Three-speed manual
Brakes:	Four-wheel drums
Body/chassis:	Separate chassis with convertible bodywork

UNITED STATES

113

FERRARI 250GT LUSSO

Lusso means luxury, and although the 250GT Lusso was not quite as refined as other supercars, it was the most indulgent of Ferrari's 250 range.

The body was steel, but the doors, hood and boot were made of alloy, and the same material was used for other parts to reduce the weight. It is believed that some all-aluminium examples were built.

Ferrari favourite Pininfarina designed the Lusso's body, although it was built by the coachbuilding firm of Scaglietti, who modified the detailing slightly for production purposes.

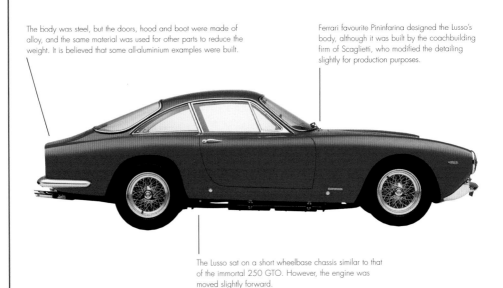

The Lusso sat on a short wheelbase chassis similar to that of the immortal 250 GTO. However, the engine was moved slightly forward.

The same V12 engine used in other 250s could be found in the Lusso, although it was detuned slightly.

Ferrari's 'luxury' machine hardly went overboard with the comfort, but improvements over previous Ferraris included more space, comprehensive and well laid out instrumentation, bucket seats and a centre console. But there was still no glovebox or room for a radio.

Instead of a full length bumper, small chrome protectors were fitted underneath the indicators. In terms of protection, they were probably quite useless, but visually they were a very attractive feature.

ITALY

115

The 250GT Lusso was the curtain call for Ferrari's long-running 250 series, and in the minds of many Ferrari fans, it was a highlight of the range. Compared to previous cars, the Lusso was more sophisticated, concerned more with grand touring than being an absolute sports machine. It marked a change in Ferrari's attitude towards its customers. Yes, the machines would still be fast, and still handle superbly. But from now on, the marque would also emphasize comfort and civility.

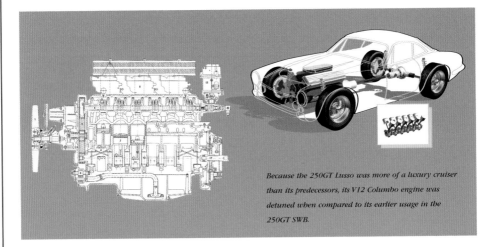

Because the 250GT Lusso was more of a luxury cruiser than its predecessors, its V12 Columbo engine was detuned when compared to its earlier usage in the 250GT SWB.

The 250 range had been born in 1953, with the 250 Europa, the first Ferrari that was intended specifically for the road. Previous models had essentially been racers that were tamed for the street. In the following years, there were several different variants of 250, including the GTO, which is generally acknowledged as the greatest Ferrari of all time.

'L' FOR LUXURY

Replacing the 250GT SWB in 1962 was the 250GT/L Berlinetta. The 'L' stood for Lusso, an indication that this sports car was more user-friendly than previous Prancing Horses.

The ante was not increased by very much, as the Lusso was still a basic machine compared to other cars that traded on their luxury status. But the concessions were there, with even the V12 engine being detuned slightly to make it more agreeable for general use. Styling was by Ferrari, and a very pretty and elegant affair it was, with its sloping roof line ending in an aerodynamically efficient Kamm-style tail.

Manufacture started in 1963, but after just one year, Ferrari felt it was out of date and replaced it with the 275GTB.

Ferrari 250GT Lusso

Top speed:	233km/h (145mph)
0–96km/h (0–60mph):	8.1 secs
Engine type:	V12
Displacement:	2953cc (180ci)
Max power:	186kW (250bhp) @ 7500rpm
Max torque:	360Nm (266lb-ft) @ 5500rpm
Weight:	1361kg (2995lbs)
Economy:	4.96km/l (14mpg)
Transmission:	Four-speed manual
Brakes:	Four-wheel discs
Body/chassis:	Separate tubular steel frame with alloy and steel two-seater coupe body

ITALY

117

FERRARI SUPERAMERICA

The name says it all. The luxurious Superamerica was built primarily with the US market in mind, and was the ultimate Ferrari of its era.

As lovely as it looks, the swooping styling of the 400 SA left little room in the boot. Ferrari therefore designed its own luggage to fit.

The 400 SA (as the Superamerica was abbreviated to) was almost exclusively bodied by Pininfarina. Just two cars were built by Scaglietti, one a Berlinetta (sedan), the other a Spyder (convertible). This one is a Pininfarina fastback coupe, the most numerous type to be made, with aluminium bodywork.

Previous big Ferraris had their V12 engines designed by Aurelio Lampredi, but the 400 Superamerica saw the debut of a new V12 by Gioacchino Colombo.

Ferraris of this era rarely carried identification badges, but the 400 SA was an exception. Its full title was spelt out on the boot, underneath the crossed flags of Ferrari and Pininfarina. One 1960 car – an unusual looking example commissioned for Fiat boss Gianni Agnelli – carried no badging at all, possibly to avoid upsetting his employees!

Battista (Pinin) Farina's personal car had retractable headlamps, but ones designed for customers had either exposed units or units that were cowled behind clear plastic.

ITALY

119

Since World War II, European sports cars have been very popular in the United States. America, though, has always demanded a bit more in the way of glitz and comfort, and these are both elements that manufacturers from across the Atlantic have sometimes been unable - or even unwilling - to supply.

The 400 Superamerica was created by Ferrari as a grand tourer to cater for American tastes. The marque's main focus had traditionally been on performance, meaning that its cars were not always the most relaxing of environments for driving. And when it came to luxury, well, Ferrari rarely did!

Previous big Ferraris had used a V12 designed by Aurelio Lampredi. For the Superamerica, this was superseded by a V12 engineered by Gioacchino Colombo, an improvement on Lampredi's engine.

SEDUCING THE STATES

The Ferrari 410 Superamerica of 1956 was a major step towards wooing the United States, but its replacement, the 400 Superamerica, took the 'user friendly' theme further. With its servo-assisted four-wheel disc brakes, overdrive as standard and Koni shock absorbers, it was a far easier car to drive than previous Prancing Horses. The V12 engine was also 'new', or rather it was a development of an existing unit that had not been used in a big Ferrari before. The inside was not the epitome of sumptuousness, but the cabin was well-trimmed and had a comprehensive range of equipment – at a high price, though.

It was available custom-bodied, but the vast majority of customers chose Pininfarina to do the coachwork, most cars appearing as extremely handsome 'Aerodinamica' coupes. Convertibles were built as well.

There were two ranges of 400 Superamericas, with the Series II of 1962 having a lengthened wheelbase, making it even more of a practical proposition. Production ceased after 54 cars in 1964, with its replacement being the visually similar 500 Superfast.

Ferrari 400 Superamerica

Top speed:	257km/h (160mph)
0–96km/h (0–60mph):	7.8 secs
Engine type:	V12
Displacement:	3967cc (242ci)
Max power:	253.5kW (340bhp) @ 7000rpm
Max torque:	352Nm (260lb-ft) @ 5500rpm
Weight:	1300kg (2850lbs)
Economy:	5.31km/l (15mpg)
Transmission:	Four-speed manual with overdrive
Brakes:	Four-wheel discs
Body/chassis:	Separate chassis with aluminium two-door coupe body

ITALY

FERRARI DAYTONA

The front-engined Daytona was already outmoded when it arrived in 1969, but that did not stop it becoming one of the best-loved Ferraris of all.

Air conditioning was only an option. If it was not fitted, these vents behind the rear window would help keep the cabin cool, as well as remove stale air from inside.

The lithe Daytona shape – whether in coupe or convertible form – was one of Pininfarina's most inspired creations, its uncluttered simplicity a major aspect of its beauty. It was designed with the aid of a wind tunnel, and the sharp nose helped immensely with the Ferrari's drag factor.

The quad-cam, all-alloy V12 engine, fitted with six twin-choke carburettors, was in the front of the car, while the transmission was at the rear. This arrangement made the Daytona's weight distribution almost perfect.

The two vents in the hood allowed hot air to escape from the engine bay.

At first, the headlamps were covered by perspex, but from 1971, retractable units were fitted, rising to reveal the bank of four lights when needed. Some cars, such as this one, also had small driving lights fitted under the wraparound indicators so that the main lights would not have to be raised when flashing other cars.

ITALY

Even for a firm like Ferrari, well-versed in building the world's most exciting and beautiful supercars, the Daytona was a thing of greatness. Unveiled when the rest of the world was turning towards mid-engined sports cars, the Daytona represented the culmination of Ferrari's front-engined experience. Everything seemed to come together with the Daytona to create a vehicle that was nothing less than magnificent.

It was Pininfarina's stunning styling that was the crowning glory of the Daytona. Not a line was out of place or ill-conceived. From

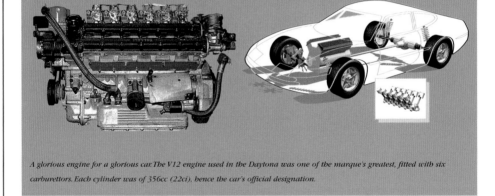

A glorious engine for a glorious car. The V12 engine used in the Daytona was one of the marque's greatest, fitted with six carburettors. Each cylinder was of 356cc (22ci), hence the car's official designation.

any viewing angle, it was a masterful opus, appearing both muscular and dramatic, yet trim and pretty at the same time.

Underneath the skin, there was little new: the chassis from the old 275 was carried over, and the familiar alloy quad cam V12 reappeared as well, albeit with more power. Yet, over the four years of its production – 1969 to 1973 – almost 1400 were sold, practically mass production for a firm like Ferrari. It was proof that the Italian manufacturer had got the formula spot on.

LE MANS CHAMPION

The Daytona was unveiled at the 1968 Paris Motor Show. Officially, it was known as the 356 GTB/4, but the press quickly dubbed it the Daytona after Ferrari's success at the 1967 American 24-hour race, and the nickname stuck.

Manufacture of both coupes and Spyder convertibles began in 1969, and Ferrari also built special competition models, versions of

which won Le Mans in 1972, 1973 and 1974. Production of the last great front-engined Ferrari came to an end in 1973.

Ferrari 356 Daytona

Top speed:	280km/h (174mph)
0–96km/h (0–60mph):	5.6 secs
Engine type:	V12
Displacement:	4390cc (268ci)
Max power:	467.5kW (352bhp) @ 7500rpm
Max torque:	447Nm (330lb-ft) @ 5500rpm
Weight:	1604.5kg (3530lbs)
Economy:	4.18km/l (11.8mpg)
Transmission:	Rear-mounted five-speed manual
Brakes:	Four-wheel vented discs
Body/chassis:	Steel square tube separate chassis with alloy and steel two-door coupe or convertible body

ITALY

125

FERRARI DINO

Enzo Ferrari marketed the mid-engined Dino as a separate marque, in tribute to his son. It became one of the best-loved Ferraris of all time.

The V6 engine was closely related to Ferrari's Formula 1 engines of the 1950s, with the cylinder dimensions being practically the same.

The door handles were small, but very elegant, the levers mounted on the tops of the doors so as not to spoil the smooth sides of the Dino.

The Dino had almost perfect balance, thanks to its compact mid-mounted V6 engine and well-designed chassis. That meant the ventilated disc brakes could be the same size on all four-wheels. It also promoted superb handling.

Up until 1969, Dinos had alloy bodywork, but this was replaced by steel bodywork for the 246GT, to make manufacture easier.

A distinctive feature of the Dino was its flying buttresses at the rear. These enveloped the engine cover, yet continued the flowing, attractive lines of the bodywork. Pininfarina's efforts on the Dino were particularly effective.

Most Dinos now wear Ferrari badges. The cars were never badged thus, but it is common for owners to add the illustrious script, just in case there are any unlikely lingering doubts about the origin of the species!

ITALY

127

Pininfarina's reputation was cemented by its work for Ferrari. Some of the most sublime cars ever created have not only worn the legendary Prancing Horse badge but also sported Pininfarina's signature script.

One of the prettiest pieces from the Italian coachbuilder was the original Dino from 1967 to 1974. The shape of this gorgeous supercar first appeared on a concept at the 1965 Paris Motor Show. It was further developed for the 1966 show season, and by

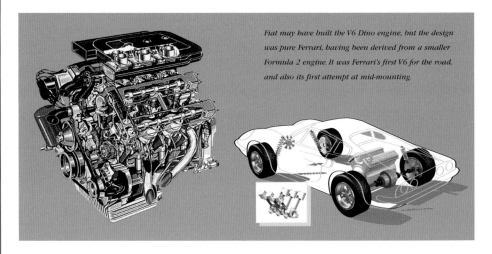

Fiat may have built the V6 Dino engine, but the design was pure Ferrari, having been derived from a smaller Formula 2 engine. It was Ferrari's first V6 for the road, and also its first attempt at mid-mounting.

1967, the definitive version had arrived, unveiled at the Turin Motor Show.

It was a Ferrari in all but name. Its identity was a tribute to Enzo Ferrari's son, Alfredino, who had died in 1957. Yet Enzo insisted on Dino being a separate marque, so no Ferrari badges were ever officially fitted. There were other departures from tradition too. It was the first mid-engined road car built by the company, and also used a V6 engine in place of the usual V12.

Because 500 had to be built to meet Formula 2 racing regulations, the engine was constructed by Fiat. However, despite having only half the usual number of cylinders, the Dino was still capable of 242km/h (150mph).

FROM ALLOY TO IRON AND STEEL

The first 206GT cars had alloy engines and bodywork. In response to complaints about mechanical reliability, an iron block was used from 1969. At the same time, the wheelbase

was extended, and steel panels were used instead of aluminium, the cars now being known as 246GTs. A targa-roofed convertible – the 246 GTS – made its debut in 1972.

Ferrari 246 Dino

Top speed:	242km/h (150mph)
0–96km/h (0–60mph):	7.3 secs
Engine type:	V6
Displacement:	2418cc (148ci)
Max power:	145kW (195bhp) @ 5000rpm
Max torque:	225Nm (166lb-ft) @ 5500rpm
Weight:	1187kg (2611lbs)
Economy:	22mpg (7.79km/l)
Transmission:	Five-speed manual
Brakes:	Four-wheel vented discs
Body/chassis:	Tubular steel chassis with steel two-door coupe body

ITALY

129

FERRARI 308

The 308 is the Ferrari that everybody remembers from the American TV series 'Magnum P.I.'. But there is much more to this Dino replacement than just small screen charisma.

The 308 was a strict two-seater, with the mid-mounted engine right up against the seats. Its positioning meant that balance was almost perfect.

Early cars had their bodywork built from fibreglass, but this was changed to steel less than two months after production started. Underneath was a traditional tubular steel chassis.

Because of the large air scoops which projected onto the doors, the handles to open them were small, delicate-looking levers mounted on top of the panels.

FERRARI 308

Aside from the four-seater Bertone-styled 308 GT4, which was a different design, there were two versions of the 308. The GTB was launched first, and had a fixed roof. The more desirable (and consequently more successful) GTS version had a removable targa roof panel.

The 308 used Ferrari's first V8 engine for the road, mid-mounted transversely. It initially featured four Weber carburettors, but from 1981, Bosch fuel injection was introduced in order to meet increasingly stringent US emissions regulations. This actually led to a fall in power, so the following year, a new *Quattrovalvole* (four-valve) head was installed, which reinvigorated the performance again.

ITALY

131

In 1973, Ferrari unveiled its replacement for the 246 Dino. The new 308 GT4 2+2 also adopted the Dino name, and was finally badged as a Ferrari (its predecessor never having been afforded that privilege).

However, the general consensus was that the new Dino was not even a patch on the original. Styled by Bertone, it had inelegant proportions, and its overall appearance was indeed clumsy.

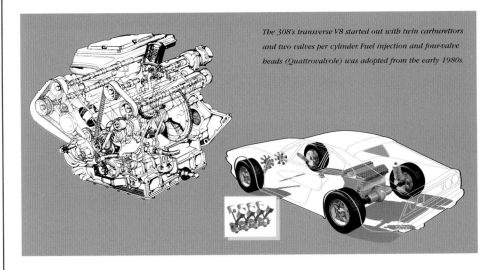

The 308's transverse V8 started out with twin carburettors and two valves per cylinder. Fuel injection and four-valve heads (Quattrovalvole) was adopted from the early 1980s.

So, in 1975, Ferrari had another go. And this time, it asked its usual collaborator, Pininfarina, to do the honours. The outcome – the two-seater 308 GTB – was far happier. In fact, it was lovely. The clean shape, while very alluring, also projected the right air of aggressiveness to make it clear to other road users that this was a Ferrari which meant business. Although its mechanicals were the same as the GT4, it had a shorter wheelbase, making it a better performer than its less appealing sister.

TAKING THE TOP OFF

A big boost for the 308 range came in 1977 at the Frankfurt Motor Show, which was the venue chosen for the launch of the 308 GTS. What made this car special was its targa roof. The roof panel could be taken off and popped in the boot (it was about all the boot could hold…), thereby making the Ferrari almost a convertible. No surprise, then, that buyers lapped it up.

In 1985, the 308 GTB/GTS line came to an end, although in name only. The body style continued on the more extreme 328 GTB and GTS. Some 6416 were built, while the 308 GT4 struggled to reach a lowly 3666.

Ferrari 308

Top speed:	233km/h (145mph)
0–96km/h (0–60mph):	7.3 secs
Engine type:	V8
Displacement:	2927cc (177ci)
Max power:	153kW (205bhp) @ 7000rpm
Max torque:	245Nm (181lb-ft) @ 5000rpm
Weight:	1502kg (3305lbs)
Economy:	5.95km/l (16.8mpg)
Transmission:	Five-speed manual
Brakes:	Four-wheel vented discs
Body/chassis:	Square-tube steel chassis with steel two-door, two-seat body with targa top

ITALY

FERRARI MONDIAL

The Mondial is now a forgotten supercar, yet its 2+2 layout was very practical, and it was the first Ferrari with fuel injection from new.

The Mondial is not regarded as one of Pininfarina's greatest Ferrari designs. The brief called for a mid-engined design but with enough room for four people inside, so it was a tough proposition even for a Ferrari master like Pininfarina. Critics felt the styling was too long, and the roof line of the coupe version too tall. Pre-1982 cars also had heavy black plastic bumpers, which looked cheap and tacky on a car such as a Ferrari.

The small, rectangular air intakes denote this car as a Mondial T. Originally, Mondials had their V8 engines mounted transversely, but in 1989, the engine was redesigned with capacity increased. It was mounted longitudinally, but the gearbox placed transversely (hence the 'T'). Prior to this change, Mondials had larger, trapezoidal-shaped grilles, which were criticized for breaking up the otherwise smooth flanks.

The radiator had to be steeply angled to fit in the front boot of the car, and there was little room for luggage. However, the Mondial also had quite a usable rear boot.

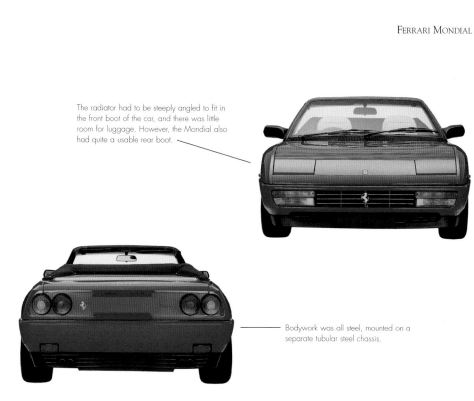

Bodywork was all steel, mounted on a separate tubular steel chassis.

ITALY

Ferrari had little luck with its V8, mid-engined 2+2s. The Bertone-designed Dino 308 GT4 of 1975 was criticized for its awkward looks, and its replacement, 1980's Mondial, was the target of similar complaints. This time, though, Bertone was not to blame. It was Pininfarina, usually beyond reproach when it came to crafting Ferraris, that had been responsible for the look of the latest 2+2 from Maranello.

To give Pininfarina its due, designing a mid-engined four seater with a large V8 situated almost in the centre was never going to be the easiest of propositions. And

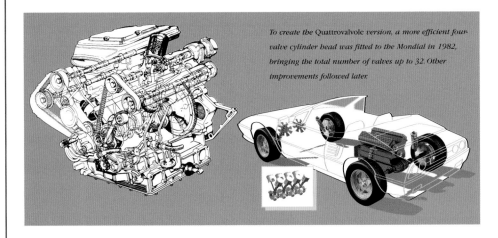

To create the Quattrovalvole *version, a more efficient four-valve cylinder head was fitted to the Mondial in 1982, bringing the total number of valves up to 32. Other improvements followed later.*

over the years, the Mondial did get better as it matured.

The Mondial was an improvement on its predecessor. It had a longer wheelbase, with more interior space inside, and as the first production Ferrari to be fitted with fuel injection from new, its power delivery was smooth. Compared to the rest of Ferrari's range, it was mild-mannered but more practical. And, like any Ferrari, if you pressed the right buttons, things could get wildly exciting very quickly.

CHANGING FACES

Mindful that the model had not been well-received, Ferrari soon introduced changes. In 1982, the ugly black bumpers were thankfully deleted, and a more efficient *Quattrovalvole* (four-valve) head was fitted. For 1983, a cabriolet version came out.

The last major alteration was in 1989, with the Mondial T. Engine power went up once more, and the V8 and gearbox were repositioned on the chassis,. There was another minor restyle too. In this more animated 'T' trim, the Mondial continued until 1994.

Ferrari Mondial

Top speed:	248km/h (154mph)
0–96km/h (0–60mph):	6.6 secs
Engine type:	V8
Displacement:	3405cc (208ci)
Max power:	224kW (300bhp) @ 7200rpm
Max torque:	321Nm (237lb-ft) @ 4200rpm
Weight:	1470kg (3235lbs)
Economy:	7.43km/l (21mpg)
Transmission:	Five-speed manual
Brakes:	Four-wheel vented discs
Body/chassis:	Separate tubular steel chassis with steel two-door 2+2 coupe or convertible body

ITALY

FERRARI TESTAROSSA

The Prancing Horse's flamboyant flagship of the 1980s was the Ferrari Testarossa, a hot-blooded V12 super machine with eye-catching Pininfarina looks that exuded pure machismo.

Mid-mounted engines are notoriously difficult to work on due to the limited space. On the Testarossa, though, the whole rear subframe could be dropped, complete with engine and gearbox in situ, to allow maintenance. That said, this was still a job best left to Ferrari professionals.

Styling, as usual, was by Pininfarina, but all the stops were pulled out to make the alloy and steel-bodied Testarossa as extravagant as possible.

The most distinctive visual aspect of the Testarossa was its side strakes, feeding air into the radiators. This theme was repeated at the back, where the rear lights also had slats covering the lenses.

Ferrari could supply specially tailored luggage to fit in the front boot – not that there was much room even with this, thanks to the sharply sloping nose and battery installation.

Needless to say, the Testarossa's cylinder heads were painted red.

The Testarossa was the widest production car of its day, at a fraction under 2m (78in). The rear track was also wider than at the front, to make up for the weight of the engine.

ITALY

139

The name means 'redhead' in Italian. But this is not an oblique tribute to some flame-haired temptress from Enzo Ferrari's past. The moniker was applied to Ferrari's 1984 V12 supercar in homage to the marque's racing cars of the 1950s, which had red-painted cylinder heads. There was actually nothing in common between the Testarossa and Ferrari's early cars except for that dash of red under the hood...and the awesome performance.

Redheads have something of a reputation for having passionate, feisty natures with a hint of unpredictability thrown in for good measure. In this sense, the Ferrari Testarossa was aptly named. An incredible-looking

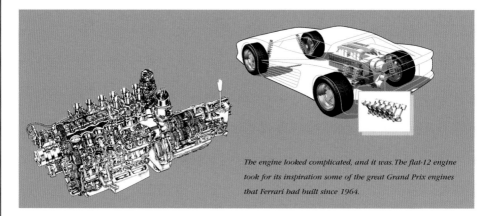

The engine looked complicated, and it was. The flat-12 engine took for its inspiration some of the great Grand Prix engines that Ferrari had built since 1964.

supercar, it oozed both charisma and attitude, and was capable of extraordinary and surprising behaviour.

EIGHTIES EXUBERANCE

The Testarossa was the successor to the Berlinetta Boxer, itself a tough act to follow. But the new 1984 Ferrari was up to the task. After the understated styling of the Boxer, Pininfarina changed tack and produced a spectacularly moody creation with strakes down the side to aid cooling. The mid-mounted V12 engine, with its tremendous exhaust soundtrack, was revised to offer – of course – more power, and handling was also enhanced thanks to the Testarossa's low stance. It took a competent driver to extract the best from the redhead, but the car was immensely rewarding no matter who was the pilot.

Eight years after its launch, power was increased and the car renamed the 512TR. Further mechanical and cosmetic changes

followed in 1994 with the 512M. The Testarossa was finally replaced in 1996 by the front-engined 550 Maranello, a more compliant but less memorable effort.

Ferrari Testarossa

Top speed:	273.5km/h (170mph)
0–96km/h (0–60mph):	5.4 secs
Engine type:	V12
Displacement:	4942cc (301.5ci)
Max power:	291kW (390bhp) @ 6300rpm
Max torque:	488Nm (360lb-ft) @ 4500rpm
Weight:	1670kg (3675lbs)
Economy:	4.96km/l (14mpg)
Transmission:	Five-speed manual with limited slip differential, mounted below engine
Brakes:	Four-wheel vented discs
Body/chassis:	Square-tube steel chassis with alloy and steel two-door coupe body

ITALY

FERRARI F40

The F40 was a back to basics supercar for Ferrari. It rejected civility to get back to doing what Ferraris always did best: going fast!

The F40 was fitted with dual fuel tanks, with the two fillers positioned on either side of the rear roof pillars.

Body was carbon fibre and Kevlar.

Pininfarina came up with a body style that reflected the F40's demeanour. It was sharp-edged and brooding, with a large dose of menace. The very obvious rear wing created considerable downforce to push the rear of the F40 down on the road. Underneath, the car had a flat underbelly, which also helped create ground effect.

For the ultimate in showing off – as if just having an F40 was not enough – the quad-cam, 32-valve, all-alloy engine can be seen through the glass in the rear hood. It may have been only a V8 of 2936cc (179ci), but the two IHI turbochargers and Marelli Weber fuel injection helped push power up to an extraordinary 356kW (478bhp). And if you wanted a bit more, a Ferrari-supplied kit could add a further 149kW (200bhp)!

The F40 had three rear pipes. Two came from the engine, while the middle one was for the turbo wastegate.

ITALY

In 1987, the Ferrari marque was 40 years old, and Enzo Ferrari suggested doing something special to celebrate its birthday. What he came up with, though, was more than just a stale party with a few balloons and some dry cake. Instead, he envisaged something a bit more extraordinary: creating the greatest Ferrari of modern times.

The F40 was a distillation of all Ferrari knew about racing and road cars in one fantastic whole. Its raison d'être was not to

With the F40 intended as a celebration model, Ferrari put great effort into building the best V8 it could. The mid-mounted engine was positioned as far forward as possible to aid balance, and each unit was tested before installation.

be the most luxurious, easiest to drive or complicated Ferrari. It was just to be the absolute fastest, best-handling creature around. Built with no compromises, it was pure power given automotive form.

FEARSOME F40

Incredibly, the car took less than 12 months to put into production, based around the tubular steel chassis of the previous 288 GTO. However, in its pursuit of sheer speed, Ferrari threw out everything that was not needed, and the F40 had a basic stripped-out interior and carbon fibre and Kevlar body panels. There were no driver aids other than what was needed to propel the car in a straight line or around corners as fast as possible.

There was nothing subtle about the noisy F40, so Pininfarina's body design was as bold and flamboyant as possible. One look at the F40, and you knew exactly what it was capable of. The V8, with its twin turbochargers and fuel injection, just proved the point.

Just 1315 F40s were built until 1992 and it is still regarded as one of the ultimate Ferraris. What a way to celebrate a birthday!

Ferrari F40

Top speed:	323km/h (201mph)
0–96km/h (0–60mph):	4.2 secs
Engine type:	V8
Displacement:	2936cc (179ci)
Max power:	356kW (478bhp) @ 7000rpm
Max torque:	573Nm (423lb-ft) @ 4000rpm
Weight:	1102kg (2425lbs)
Economy:	8.50km/l (24mpg)
Transmission:	Five-speed manual, LSD
Brakes:	Four-wheel vented discs
Body/chassis:	Carbon fibre and Kevlar body panels with welded steel tube cage and suspension mounts

ITALY

Ferrari 456GT

The 456GT does more than just echo the classic Daytona with its alluring styling. It has also reverted to having a V12 engine at the front.

The rear tyres are slightly wider than those at the front, as they have to deal with all the stress of transmitting that V12 power onto the road.

The body is made of alloy, as are some of the mechanical parts, including the engine. Despite this, it is not exactly lightweight. The tubular steel chassis is heavy, and all the extra sophisticated equipment fitted to the Ferrari means it weighs in at over 1818kg (4000lbs)

The ZF transmission is mounted at the back, to balance out the hefty V12 at the front.

The cabin is so airtight, it can be difficult to close the doors. So, when the door is shut, the window opens slightly to reduce air pressure, then closes again once the car is locked.

Compare the styling to that of the 1968–74 Ferrari Daytona, and Pininfarina's inspiration for the 456 GT suddenly becomes very obvious. It is most noticeable from the rear, where the four round lights, the curve of the boot and the angle of its aperture are almost modern copies of the Daytona.

ITALY

According to Ferrari itself, the 456GT was the marque's first car to be designed with the 21st century in mind. If this is true, it was a surprisingly backward-looking car. For not only did it feature a front-mounted V12 engine for the first time in almost 20 years, but it also used the last great car where that configuration had appeared – the celebrated Daytona – as the inspiration for its graceful styling.

Throughout the 1970s and 1980s, Ferrari's flagship V12s had all featured mid-mounted

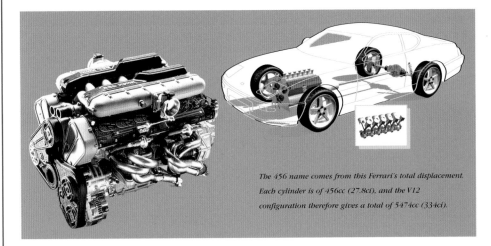

The 456 name comes from this Ferrari's total displacement. Each cylinder is of 456cc (27.8ci), and the V12 configuration therefore gives a total of 5474cc (334ci).

engines. In 1988, just before he died, Enzo Ferrari saw work start on the car that would return to the older values of the 1960s and before. Suspension technology had moved on since then, and with electronic damping and effective power steering, it was now possible to build a front-engined car that handled just as well as something with its engine in the centre.

DAYTONA HERITAGE

The 456 GT 2+2 was launched at 1992's Paris Motor Show, to instant acclaim. Pininfarina's styling was a triumph, managing to conjure up visions of the old Daytona, but still looking fresh and modern enough for the forthcoming new millennium. The formidable quad-cam V12 was new, and with a six-speed ZF transaxle at the rear, balance was almost perfect. That gave the 456GT impeccable road manners, making it among the easiest of all Ferraris to drive, though it was still able to hit phenomenal velocities.

In 1996, the GT was complemented by the GTA, with a four-speed automatic gearbox, which made it even more effortless to drive. Manufacture of both models continues to this day.

Ferrari 456GT

Top speed:	299km/h (186mph)
0–96km/h (0–60mph):	5.2 secs
Engine type:	V12
Displacement:	5474cc (334ci)
Max power:	324kW (435bhp) @ 6250rpm
Max torque:	550Nm (406lb-ft) @ 4500rpm
Weight:	1824kg (4013lbs)
Economy:	4.18km/l (11.8mpg)
Transmission:	Rear mounted six-speed manual or four-speed automatic
Brakes:	Four-wheel vented discs
Body/chassis:	Tubular steel chassis with two-door 2+2 alloy bodywork

ITALY

149

FERRARI 355

It may have been only an 'entry level' Ferrari, less demanding than most Italian supercars, but in looks and performance, the F355 was still absolute bliss.

The Pininfarina signature badge on the rear flank testifies that this is yet another great design by Ferrari's main styling collaborator. Both coupe and cabriolet versions were made.

In keeping with Ferrari tradition, the F355 had a separate chassis, albeit one with a mix of tubular and sheet steel. The open-top Spider version had a stiffer chassis, but despite the extra metalwork, it was no heavier than the coupe model.

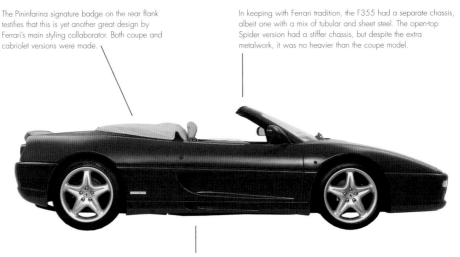

The F355 was an evolution of the previous 348 range, and the styling was almost identical. One of the most noticeable visual differences could be found down the side of the car, where the straked air scoops were replaced by one slightly more subtle but larger opening.

The mid-mounted all-alloy V8 was mounted longitudinally instead of transversely, the same as in the preceding 348 range, which was the first Ferrari to feature this arrangement. The four overhead cams operated five valves per cylinder, three intake and two exhaust.

A smooth undertray was fitted to improve the aerodynamics, as well as help generate ground effect, helping the F355 stick to the road.

ITALY

In 1994, Ferrari unveiled the latest in its family of V8 cars. The line had begun in 1973 with the 308GTB, each subsequent model bringing something new to the mix. However, even by Ferrari standards, the F355 was a big step forwards, threatening to eclipse its bigger V12 Testarossa sister. For all their bravado and beauty, Ferraris have never been known as easy cars to drive. Even the most-cherished examples have always demanded a lot of effort from their drivers, and while the rewards have invariably been

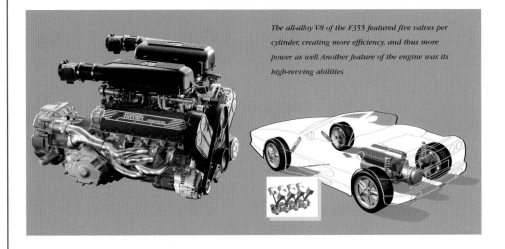

The all-alloy V8 of the F355 featured five valves per cylinder, creating more efficiency, and thus more power as well. Another feature of the engine was its high-revving abilities.

worth it, a Ferrari is not usually a car to relax in. The F355 changed all that. At last, here was a Ferrari that was refined, comfortable, easy to drive, and at home with any road conditions. It came with an unexpected level of luxury and sophistication, including air-conditioning, power steering, ABS and a great-sounding stereo.

Yet it also displayed all the usual Prancing Horse characteristics of ferocious speed, intense power and superlative roadholding.

CHANGES FOR THE BETTER

The F355 used the same chassis as the previous 348 and the V8 engine was only slightly modified. Elsewhere, however, there were enough changes to make the F355 a big improvement over its immediate ancestor. And the new six-speed manual gearbox made it user-friendly.

Pininfarina's styling, a variation on the 348 theme, was understated, avoiding the fussy extravagances that had spoiled other

Ferraris. The whole package was sheer brilliance.

The F355 was replaced by the new 360 Modena in 1999, which managed to increase standards even more.

Ferrari 355

Top speed:	294.5km/h (183mph)
0–96km/h (0–60mph):	4.7 secs
Engine type:	V8
Displacement:	3496cc (213ci)
Max power:	279kW (375bhp) @ 8250rpm
Max torque:	363Nm (268lb-ft) @ 6000rpm
Weight:	1353kg (2977lbs)
Economy:	6.4 km/l (18.1mpg)
Transmission:	Six-speed manual
Brakes:	Four-wheel vented discs, ABS
Body/chassis:	Tubular and sheet steel central chassis with alloy two-door, two-seat coupe or convertible body

ITALY

FERRARI 550 MARANELLO

Christened after Ferrari's home town, the 550 Maranello was an Italian supercar that combined practicality and driving ease with epic performance and superior handling.

Pininfarina did the styling of the 550 Maranello, but it was not hailed as one of its masterpieces. Although attractive, it did not stand comparison with the styling house's classic Ferraris, being labelled as not distinctive enough and too fussy in places.

The V12 mounted at the front was a return to 1960s values, and all the big Ferraris after this decade had their engines mid-mounted. The V12 was the same as used in the 456 GT, but engineers were not content to leave it untouched, and managed to force another 36.5kW (49bhp) out of the unit.

The 550 Maranello used a traditional floor-mounted, alloy-gated gear shifter. However, its current manifestation, the 575 Maranello, is the first V12 Ferrari road car to use a Formula 1 style transmission, with paddles on the steering wheel.

The body was made of aluminium, mounted on a steel frame.

The cabin of the 550 Maranello was well-equipped, especially by Ferrari standards. There was air-conditioning, leather, seats that would could be power-adjusted eight ways, and a CD player.

Round tail lamps are a long-running Ferrari trademark.

When Ferrari stopped building front-engined V12 two-seater supercars back in the 1970s, this was because it felt mid-engined designs offered the way ahead in terms of handling. And so they did, for a while. An engine in the middle offers the best balance for a car, although it is also a compromise because cabin space is limited by the intrusion of the engine.

Automotive technology changed vastly between the 1970s and the 1990s, though,

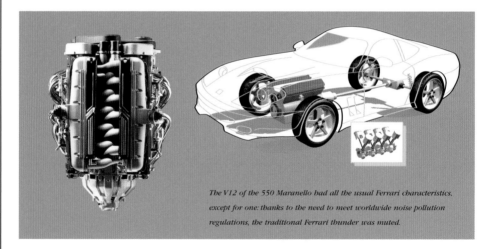

The V12 of the 550 Maranello had all the usual Ferrari characteristics, except for one: thanks to the need to meet worldwide noise pollution regulations, the traditional Ferrari thunder was muted.

and by 1992 Ferrari felt it could at last engineer a front-engined car that could surpass a mid-engined one. That was the 456GT, and it marked a turning point.

BACK TO FRONT

The 456 GT was a four-seater. Fans and customers had to wait until 1996 for the first two-seater front-engined Ferrari V12 in years. At its Fiorano test track launch, Michael Schumacher put the 550 Maranello through its paces, but even without his presence, the 550 would have captured headlines.

It was the fastest street Ferrari then available, with significantly increased performance over the previous 512M. And it was also a more refined car, easier to live with, and more forgiving in action. For those who wanted Ferrari standards but did not want to stint on sophistication, the 550 was the one to choose.

Only the innocuous styling disappointed. It simply was not Ferrari-like. Had critics

been able to sign Pininfarina's report card for the 550 Maranello, most would have put, 'Could try harder'.

The 550 grew into the modified 575M Maranello in the twenty-first century, with a bigger engine, and cosmetic improvements.

Ferrari 550 Maranello

Top speed:	320km/h (199mph)
0–96km/h (0–60mph):	4.4 secs
Engine type:	V12
Displacement:	5474cc (334ci)
Max power:	362kW (485bhp) @ 7000rpm
Max torque:	539Nm (398lb-ft) @ 5000rpm
Weight:	1694kg (3726lbs)
Economy:	4.18km/l (11.8mpg)
Transmission:	Six-speed manual
Brakes:	Four-wheel vented discs
Body/chassis:	Steel frame with two-door aluminium coupe body

ITALY

157

FERRARI 360 MODENA

Every new Ferrari improves on the ones before it, and the current mid-engined Modena is a technical and visual marvel, setting new standards even for Ferrari.

The 360 makes do without any spoilers to generate downforce, yet still develops more than in the previous 355 model. It took a lot of wind-tunnel testing for Pininfarina to come up with a shape that was both this effective and attractive.

A first for Ferrari was the use of all-aluminium construction for the 360. It had built cars with alloy bodies before, but the monocoque spaceframe and even the suspension components are aluminium on the Modena.

Ferrari's V8 was completely redesigned for the 360. It features 40 valves (three intake and two exhaust per cylinder), and four camshafts, two for each cylinder bank. Parts inside the engine are made from tough forged aluminium and titanium. It is a very high-revving unit, with maximum power generated at 8500 revolutions per minute.

Previous V8 Ferraris had their radiators mounted alongside the engine, but in the Modena, they are mounted in the nose, one in each corner.

A brand new six-speed manual gearbox made its first appearance in the 360.

ITALY

For Ferrari, replacing its F355 was not an easy proposition. The V8-engined F355 had been an exceptional car, and anything intended to usurp it would have to be remarkable indeed.

However, the 360 Modena of 1999 not only succeeded in this brief but also offered unprecedented levels of performance. Indeed, it surprised even those who expected great things from the Prancing Horse. Named after the location of Ferrari's factory, it was described by the company as a 'clean sheet design anticipating trends for future Ferrari road cars'.

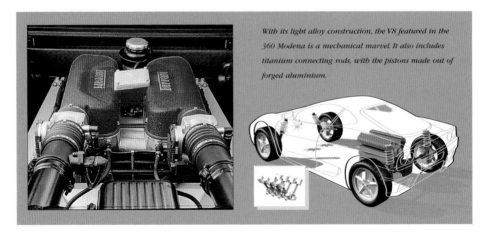

With its light alloy construction, the V8 featured in the 360 Modena is a mechanical marvel. It also includes titanium connecting rods, with the pistons made out of forged aluminium.

ALUMINIUM REVOLUTION

Even for a Ferrari, it was pretty radical. The high-backed styling was svelte and beautiful, a flowing masterpiece by Pininfarina that was also incredibly aerodynamic. Just as fascinating as the shape was the engineering. It was of monocoque construction, but completely built out of aluminium. Ferrari had never attempted such a thing before. The body, spaceframe chassis, suspension components and transmission casing were all built out of the lightweight material, a factor which significantly enhanced handling and performance.

The V8 engine was carried over from the F355, but reworked to give - surprise, surprise - more power. For its size, it gave an astonishingly high output, with each litre of the engine developing 82kW (111bhp), a statistic which put it among the very best of performance engines.

Thanks to its fitted traction control, handling was nothing short of unbelievable, and the overall impression of the Modena as one of the most civilized Ferraris ever built was further fortified by the spacious and comfortable interior. It continues to be built to this day.

Ferrari 360 Modena

Top speed:	298km/h (185mph)
0–96km/h (0–60mph):	4.5 secs
Engine type:	V8
Displacement:	3586cc (219ci)
Max power:	294kW (394bhp) @ 8500rpm
Max torque:	372.5Nm (275lb-ft) @ 4750rpm
Weight:	1393kg (3065lbs)
Economy:	5.66km/l (16mpg)
Transmission:	Six-speed manual
Brakes:	Four-wheel vented discs
Body/chassis:	Aluminium spaceframe/monocoque

ITALY

FIAT DINO

Fancy a convertible Italian sports car with a famous name and Ferrari engine but think you cannot afford it? Then try the Fiat Dino.

When engine size went up in 1969 – from 1987cc (121ci) to 2418cc (147.5ci) – Fiat fitted a ZF five-speed manual gearbox in place of the original unit, which proved unable to handle the extra power. That had been a 2300 four-speed box, although modified with a separate casing to make it a five-speed.

Just as it had styled Ferrari's version of the Dino, Pininfarina also did Fiat's convertible Spider variant. However, the hard top coupe – which appeared five months after the Spider – was done by Bertone, and although very attractive in its own right, its looks did not quite reach the standard of its open-air sister.

Conceived by Ferrari, the Dino's quad-cam V6 engine, which could map its heritage back to the 1950s, was engineered by Aurelio Lampredi.

Fiat raided its parts bin to make the Dino. The circular rear lights were from the 850 coupe, the floorpan, rear suspension and gearbox came from the 2300S coupe, while the 124 Spider donated its front suspension. Later on in its life, back suspension from the 130 sedan was adopted.

During the mid-1960s, the rules for Formula 2 racing changed. The new guidelines stated that 500 examples of an engine had to be built before it would be allowed to race in the championship. This presented Ferrari with a dilemma. It wanted to go on using its V6 Dino racing engine, but as a small volume manufacturer, it had no chance of manufacturing the number of units required.

So, Enzo Ferrari went to Fiat to talk about a possible collaboration. Ferrari would build

Under the Fiat skin beat a Ferrari heart. The Dino V6 was one of the best-loved of all Italian high-performance engines, although purists preferred the earlier all-aluminium versions, even though they were less powerful.

a Dino-engined car, so would Fiat. Between the two of them, they would easily be able to produce the required number.

THE 'OTHER' DINO

It was no Ferrari in looks, but the Pininfarina-bodied Fiat convertible unveiled at the 1966 Turin Motor Show was nevertheless very striking. And although it borrowed bits and pieces from lots of less able Fiats, the whole was greater than the sum of its parts, thanks to that splendid V6.

In 1967, it was joined by the less spectacular Dino coupe, styled by Bertone, with a longer wheelbase and four seats. As with the Ferrari Dino, power was increased in 1969 when an enlarged V6 was adopted. Simultaneously, the cars got a more robust transmission and independent rear suspension.

The last cars were sold in 1973, by which time Fiat had bought Ferrari. Although more Fiat Dinos were built than Ferrari ones, it is the latter that has the better survival rate. This is not just a legacy of Fiat's well-known rust problems, but also a result of Ferrari owners robbing Fiats to keep their own more exotic cars running!

Fiat Dino

Top speed:	209km/h (130mph)
0–96km/h (0–60mph):	7.7 secs
Engine type:	V6
Displacement:	2418cc (147.5ci)
Max power:	134kW (180bhp) @ 6600rpm
Max torque:	215Nm (159lb-ft) @ 4600rpm
Weight:	1172kg (2579lbs)
Economy:	6.37km/l (18mpg)
Transmission:	Five-speed manual
Brakes:	Four-wheel Girling discs
Body/chassis:	Unitary steel chassis with Pininfarina convertible or Bertone coupe bodywork

ITALY

FORD SHELBY MUSTANG GT500

Racer Carroll Shelby breathed fire into Ford's Mustang to create the ferocious GT500, which ranks alongside his Cobra as one of America's most evocative sports cars.

As with the Cobra, the GT500 was powered by a 428 Police Interceptor V8. Its power output was advertised at 250kW (335bhp), but this was really a way of getting around insurance company premiums. The actual figure was closer to 298kW (400bhp).

A roll cage came as standard on '69 Shelbys. Not just a safety device, it also stiffened the shell.

Locking pins were a standard fitting to anchor the fibreglass hood. Shelby Mustangs also had completely reworked front ends, which were 77mm (3in) longer than standard fenders.

The exhaust pipes were routed through the rear valance. Although this tidied up the back appearance, it did little for corrosion proofing. A more worrying proposition was the possibility of fire, as the pipes ran close to the gas tank.

Fitted at the rear of 1969–70 GT500s were sequential tail lights from the Ford Thunderbird. When the brakes were applied, the lights would flash in sequence, from the centre to the corners. They also served as indicators. Above everything else, though, they looked extremely cool.

Ford's Mustang - the original Pony Car - is an American legend. But all good legends have a hero, and in the case of the Mustang, his name was Carroll Shelby.

Texan racing driver and chicken farmer Carroll Shelby had been responsible for unleashing the overwhelming Cobra back in 1962. When Ford unveiled its 'personal coupe' two years later, Shelby was quick to get involved. Fastback bodies were shipped to his Los Angeles factory from 1965 onwards, where Shelby fitted them with

The GT500's engine was more about brute strength than engineering sophistication. Its cast-iron Ford 428 V8 was strengthened throughout to cope with the extra stresses imposed on it by Carroll Shelby.

more brawny V8 engines, carried out other mechanical and cosmetic changes, and then sold them to the public as GT350s.

In 1967 came the ultimate Shelby Mustang. The GT500 had a colossal Ford 428 V8, which pumped out so much power that Shelby had to lie about its output so that insurance companies would accept them!

SHELBY BATTLES THE BOSS

When the stock Mustang was rebodied in 1969, the Shelby followed suit, although it kept its unique front end, with lightweight fibreglass hood. However, Ford itself was by now playing the performance game and its cheaper but almost as maniacal Mach 1 and Boss models were luring sales away. Despite the GT500KR (King of the Road) special edition, production ceased in the same year, with the last sales in 1970.

For those unable to afford the real thing on a permanent basis, Hertz rented out Shelby GTs for the weekend. It was surprising just how many of these were returned with worn tyres, a few dents and signs of racing numbers having been applied to the bodywork.

Ford Shelby Mustang GT500

Top speed:	209km/h (130mph)
0–96km/h (0–60mph):	5.5 secs
Engine type:	V8
Displacement:	428ci (7014cc)
Max power:	250kW (335bhp) @ 5200rpm
Max torque:	596Nm (440lb-ft) @ 3400rpm
Weight:	1409kg (3100lbs)
Economy:	2.83km/l (8mpg)
Transmission:	Select-shift C6 Cruise-O-Matic four-speed manual or four-speed automatic
Brakes:	Discs at front, drums at rear
Body/chassis:	Unitary chassis with steel two-door fastback body

UNITED STATES

FORD RS200

Some 200 examples of the fearsome Ford RS200 Group B rally car – a rare example of a Blue Oval supercar – entered private hands as homologation specials.

This pod housed the intercooler, which fed air to the turbocharger. Although it looked unaerodynamic, its positioning was the result of wind-tunnel testing.

The windscreen came from the then current Ford Sierra sedan, while the Cosworth engine was a development of the Cosworth engine found in the Escort RS1700T. It featured a Garrett AiResearch turbocharger, plus electronic engine management.

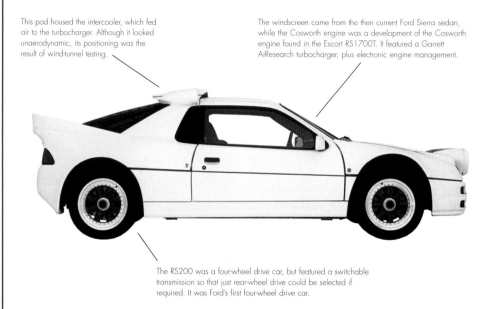

The RS200 was a four-wheel drive car, but featured a switchable transmission so that just rear-wheel drive could be selected if required. It was Ford's first four-wheel drive car.

Unlike some Group B rally cars, which bore a superficial resemblance to existing models, the RS200 was a one-off design by the Ford-owned Ghia styling studio of Italy. The monocoque shell was constructed using carbonfibre and Kevlar over a honeycomb floor.

The four-cylinder engine was mid-mounted to aid handling balance. The whole rear section of the car could be hinged upwards (or even removed completely) for ease of maintenance.

Interiors were surprisingly refined, with a fully-equipped dashboard, carpeting, and even space for a radio, not something you got with most sold-off rally cars!

RS 200

UNITED STATES

171

Ford enjoyed an enviable success record in rallying throughout the 1970s and 80s. However, its many victories came from reworked versions of road cars such as the various incarnations of the all-conquering rear-wheel drive Escort.

Something a bit more extraordinary was needed if Ford was to compete in the highly demanding Group B rally category. Ford's response was to build a rally supercar, the fastest, most maniacal Blue Oval to appear since the GT40 of almost 20 years earlier.

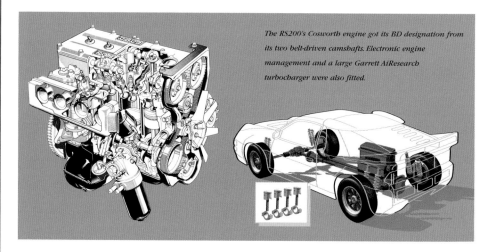

The RS200's Cosworth engine got its BD designation from its two belt-driven camshafts. Electronic engine management and a large Garrett AiResearch turbocharger were also fitted.

The RS200 lacked the sinuous looks of the GT40. In fact, its appearance was somewhat short and rotund. But it was designed not for race tracks but rather the loose gravel and ice of the rally stage. And under such conditions, it was devastating.

GHIA'S RALLYE SPORT WONDER

Launched in 1984, under the Rallye Sport (RS) name, the new Ford had a mid-mounted four-cylinder engine, Ghia-designed special lightweight body, and four-wheel drive. Despite its blocky aerodynamics, it was extremely fast, and on corners, there was little to compete with its superb road-holding ability.

Before it could enter Group B rallying, though, 200 homologation cars had to be built. Delays in production meant that this total was achieved only in early 1986, and the hold-up proved to be bad news. Ford had just won its first major events when Group B was banned after a series of horrendous crashes.

This left the RS200 without a series to compete in.

The remaining RS200s were sold off as private vehicles, making it one of the rarest and most unusual supercars.

Ford RS200

Top speed:	225km/h (140mph)
0–96km/h (0–60mph):	6.1 secs
Engine type:	In-line four
Displacement:	1803cc (110ci)
Max power:	186kW (250bhp) @ 6500rpm
Max torque:	291Nm (215lb-ft) @ 4000rpm
Weight:	1185kg (2607lbs)
Economy:	5.66km/l (16mpg)
Transmission:	Five-speed manual
Brakes:	Four-wheel vented discs
Body/chassis:	Platform-type chassis with aluminium honeycomb, carbonfibre and steel body

UNITED STATES

173

GHIA 450SS

Few people have even heard of the 450SS, fewer still have seen one. This rarest of supercars was a blend of Italian style and American muscle.

The styling of the Ghia 450SS was a curious mix of Italian and American idioms. From the front and rear, it looked like an archetypical European sports car, yet in profile, it was almost pure Yankee muscle. Giorgetti Giugiaro was the stylist, during his brief spell at Ghia. The look was inspired by Ghia-bodied Fiat 2300S, although modified to take a Plymouth chassis.

The construction was purely conventional, a steel-mounted body on a separate chassis, with the engine driving the rear wheels. The V8 was courtesy of the Plymouth Barracuda, although it was a Commando high performance version with much more power than most standard production models got. The 'Cuda also donated its chassis, which meant soft suspension and lots of body roll, taking the edge off the Ghia's handling.

As well as the more common convertible, a hard-top coupe version of the Ghia was available as well. The best compromise, though, was to turn an open-air car into something more weather-resistant by fitting the optional removable roof.

The front bumpers blending deftly into the fenders was a neat touch.

ITALY

175

Check motoring books for the Ghia 450SS, and you will be lucky to come across even a passing reference to it. Search the Internet, and you will find few websites that even mention it. Discuss the car with motoring enthusiasts, and very few will have any idea of what you are talking about. The Ghia 450SS is a forgotten supercar, in existence for only a few years before it seemed just to evaporate.

Turin-based Ghia was once one of the finer Italian coachbuilders, and before it

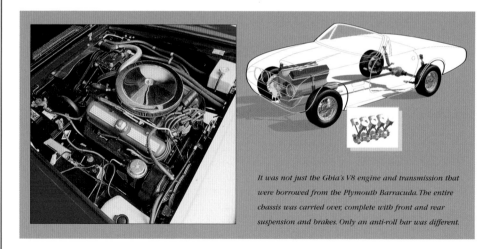

It was not just the Ghia's V8 engine and transmission that were borrowed from the Plymouth Barracuda. The entire chassis was carried over, complete with front and rear suspension and brakes. Only an anti-roll bar was different.

became a Ford subsidiary, it had strong links with Chrysler. It was a Fiat that led to the 450SS though.

SUGARMAN'S DREAM

The story goes that a Los Angeles businessman, Burt Sugarman, saw a Ghia-bodied Fiat 2300 coupe on a magazine cover and decided it would make a great sports car for the United States. It would need something a bit more special under the bonnet, of course, and this being the America of the 1960s, only a V8 would do.

He came up with the idea of the lovely Guigiaro-designed Ghia body mounted on a Plymouth Barracuda chassis complete with engine. Ghia liked the concept, and agreed to build the shells for shipping to the States. The 1966 Italian/American collaboration was four times more expensive than the 'Cuda it was based on.

However, things started to unravel almost immediately. Ghia was sold to De Tomaso in

1967, and subsequent internal ructions caused the demise of the 450SS project after just 40 had been made. A sad end for a supercar with great potential.

Ghia 450SS

Top speed:	201km/h (125mph)
0–96km/h (0–60mph):	8.5 secs
Engine type:	V8
Displacement:	4490cc (273ci)
Max power:	175kW (235bhp) @ 5200rpm
Max torque:	379Nm (280lb-ft) @ 4000rpm
Weight:	1532kg (3370lbs)
Economy:	5.66km/l (16mpg)
Transmission:	Three-speed automatic or four-speed manual
Brakes:	Discs at front, drums at rear
Body/chassis:	Separate chassis with steel two-door coupe or convertible body

ITALY

HONDA NSX

Japan's first full-blown supercar was the NSX, a car that blended typical Far East reliability and driving ease with Italian standards of performance and handling.

The mid-mounted alloy V6, with its four camshafts driven by belts, was installed transversely. The secret to its impressive performance was VTEC (variable valve timing), which improved response at higher revs, without sacrificing bottom end tractability.

The NSX used all-aluminium monocoque construction, which had the twin benefits of being extremely strong, but also very light. Three different widths of metal were used. Almost a decade had to pass before Ferrari got around to using this type of build process!

Because of the large glass area, visibility in the NSX was excellent, not something you could say about your average supercar. Later cars had the option of a removable targa roof.

Acceleration was controlled electronically, there being no mechanical linkage between the pedal and the engine.

Initially, five-speed manual and (this being a Japanese car) four-speed automatic transmissions were offered. In 1997, a six-speed manual gear box made the car even more flexible.

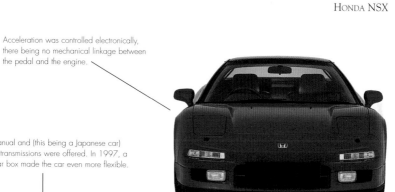

Power steering was variable. Driven by an electric motor, it gave a lot of assistance at low speeds, but this would gradually reduce the faster the car went.

JAPAN

179

It was only a matter of time before Japan came up with a supercar to take on the best that Europe had to offer. Perhaps the real surprise was that the country's automotive manufacturers had not tried to build one long before.

The mould was broken by Honda. Its NSX (New Sports Car eXperimental) project got under way in 1984, and was unveiled at the Geneva Motor Show in 1989. This long development time was justified by what finally appeared. Honda had taken the best

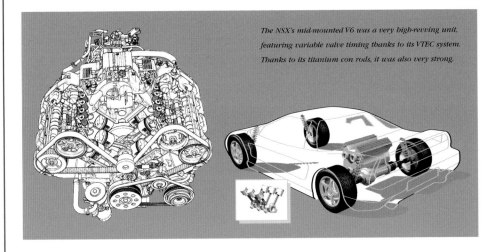

The NSX's mid-mounted V6 was a very high-revving unit, featuring variable valve timing thanks to its VTEC system. Thanks to its titanium con rods, it was also very strong.

ideas from European design to use in its own car, but had added the typical Japanese virtues of reliability, practicality and versatility. It had even asked racing driver Ayrton Senna to test the car throughout its evolution. The result was that most rare of creatures, a supercar that was just as happy ambling around town as it was at speed.

ORIENT EXPRESS

Critics claimed that the polite and forgiving NSX lacked the passion and character of a Ferrari or Lamborghini. But the simple truth was that Honda had built a damn fine supercar, free from the idiosyncrasies and faults that so often afflicted such machines. Performance was more than adequate, and handling was formidable, thanks to traction control fitted as standard. Then, in 1997, a bigger engine and six-speed gearbox made it even better.

It was sold under the Acura name in the USA, where the NSX's biggest fault was its lack of heritage. Supercar buyers preferred something with a more illustrious badge, even if it was more expensive and not as usable. As Honda found out, life can be unfair.

Honda NSX

Top speed:	261km/h (162mph)
0–96km/h (0–60mph):	5.4 secs
Engine type:	V6
Displacement:	2977cc (182ci)
Max power:	204kW (274bhp) @ 7000rpm
Max torque:	284.5Nm (210lb-ft) @ 5300rpm
Weight:	1373kg (3021lbs)
Economy:	5.73km/l (16.2mpg)
Transmission:	Five or six-speed manual/four-speed automatic
Brakes:	Four-wheel vented discs, ABS
Body/chassis:	Alloy two-door, two-seat coupe with alloy monocoque chassis

JAPAN

ISO GRIFO

The best-known Iso model is the Grifo, an Italian supercar that chose to take on Ferrari and Lamborghini using a bit of American V8 muscle.

As supercars go, the Grifo was one of the more practical examples. As well as a usable boot, it had a very large rear window (which looked like a hatchback, but was not) and this afforded superb visibility. It had its own built-in defroster too.

Instead of building its own engines, Iso used off-the-shelf V8s from the Chevrolet Corvette. The first Grifos used a small-block 327 Chevy engine of 5359cc (327ci). Then, in 1968, came a big-block Chevy 427 of 6997cc (427ci). These '7-Litre' Isos could be identified by their very obvious hood bulge.

The Grifo shape was yet another design by Giorgetto Giugiaro. It was one of his early ones, though, created while he was working for Bertone.

Air-conditioning was an optional extra inside. Putting the accent on luxury, the Grifo also had leather upholstery and a wooden dash. However, lack of interior space meant that the radio had to be positioned on the passenger's side of the dashboard.

In 1970, the front was reshaped with semi-retractable quad headlamps set in the radiator air scoop.

ITALY

Iso's entry into the world of auto manufacturing could not have been more different to the cars it would end up building. The Milanese fridge and motor scooter firm started life producing Isetta bubble cars during the 1950s, but in 1962 had a radical change of mind, and moved into the world of exotic sports cars instead.

Its first offering was the Rivolta of 1962, but within a year, it had something else to completely eclipse this car. The Grifo debuted on the supercar stage at the 1963

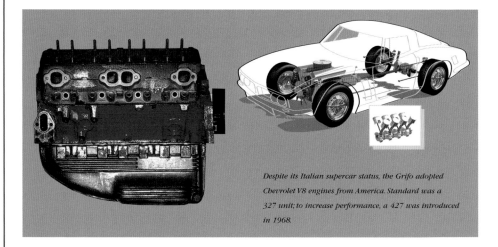

Despite its Italian supercar status, the Grifo adopted Chevrolet V8 engines from America. Standard was a 327 unit; to increase performance, a 427 was introduced in 1968.

Turin Motor Show, in both road and competition trim. With smooth and masterful styling by Giorgetto Giugiaro, and four headlamps peering out from its purposeful snout, the new Iso was a handsome beast.

It sounded impressive and drove powerfully too, thanks to its Chevrolet V8 engine (as used in the Corvette) mated to a shortened Rivolta chassis with all-round independent suspension.

POWER UP!

Grifos started reaching customers in 1965, and those fortunate and rich enough to buy one could specify either 224kW (300bhp) or 272kW (365bhp) power outputs. This was surpassed in 1968 by the 7-Litre model, which pumped out 290kW (390bhp) and was capable of taking on the best that Ferrari and Maserati could muster.

In 1970, the car had a mild facelift, which masked the headlamps behind semi-retractable panels. It was a small touch, but

one that freshened up the look ready for the new decade.

In 1974, just halfway through that decade, Grifo and its Iso were no more, both victims of the fuel crisis.

Iso Grifo

Top speed:	262km/h (163mph)
0–96km/h (0–60mph):	6.4 secs
Engine type:	V8
Displacement:	5359cc (327ci)
Max power:	272kW (365bhp) @ 5800rpm
Max torque:	488Nm (360lb-ft) @ 3600rpm
Weight:	1380kg (3036lbs)
Economy:	4.89km/l (13.8mpg)
Transmission:	ZF Five-speed manual
Brakes:	Four-wheel servo assisted disc brakes
Body/chassis:	Monocoque two-door coupe

ITALY

185

JAGUAR D-TYPE

The D-type was a purebred racer, intended to win Le Mans for Jaguar. Its offshoot was the XKSS, intended to bring D-type performance to the road.

Behind the driver's seat was a large stabilizing fin, intended to provide steadiness at high speed. It also provided a handy flat surface for displaying racing numbers!

Jaguar aerodynamicist Malcolm Sayer designed both the D-type and the later E-type, with the slippery shape of the D-type being aided by wind-tunnel tests. There were many similarities in styling between the D-type and the E-type.

The body construction was a stressed skin magnesium central monocoque, with a separate subframe holding the engine and independent front suspension. This gave the D-type great rigidity.

The D-type's six-cylinder XK engine was basically the same as found in Jaguar's production road cars, although it was of dry sump construction, with two oil pumps circulating oil around the engine. A dry sump meant the height of the engine could be reduced, and also helped prevent oil surges during enthusiastic cornering.

For racing purposes, a cover could be placed over the redundant passenger seat.

This small hatch at the rear holds the spare wheel...and very little else. D-types were not known for their luggage capacity.

UNITED KINGDOM

187

Jaguar's first win at the Le Mans 24-hour race came in 1951, when an XK120C – soon to be known as the C-type – took the laurels after a hard-fought battle. The 1953 event saw further accomplishments, C-types coming first, second and fourth.

Firmly committed to its racing programme and maintaining its success, Jaguar produced a prototype for the C-type's successor in May 1954. The logically named D-type was an outright race car, developed for the track and using advanced semi-monocoque

What else would be under the bonnet of a 1950's Jaguar than an XK engine. Developments for racing included a dry sump, more efficient exhaust and wider camshaft. Three Weber carbs were fitted.

construction plus a beautifully sculptured body designed by former aircraft engineer Malcolm Sayer. His aviation experience led him to develop the streamlined shape in a wind tunnel. There were few manufacturers employing such techniques in the 1950s.

LE MANS DOMINATION

Production began in 1955, although in strictly limited numbers, the car being intended for private racing teams. The car proved its integrity by winning the 1955, 1956 and 1957 Le Mans, the latter two victories under the flag of the Ecurie Ecosse Scottish racing team – proving the D-type was such an accomplished racer that it succeeded even without the official backing of Jaguar.

So triumphant was the D-type that Jaguar produced a raw, road-going version, dubbed the XKSS, in 1957. Effectively, this was the racer given some subtle tweaks to make it street-legal. Among the changes were bumpers, a proper windscreen and a primitive soft top.

Only 16 were built before a disastrous fire damaged Jaguar's factory, halting production. It never restarted, although elements of the XKSS lived on in the E-type, four years later.

Jaguar D-type

Top speed:	261km/h (162mph)
0–96km/h (0–60mph):	5.4 secs
Engine type:	In-line six
Displacement:	3442cc (210ci)
Max power:	186kW (250bhp) @ 6000rpm
Max torque:	328Nm (242lb-ft) @ 4000rpm
Weight:	1118kg (2460lbs)
Economy:	7.08km/l (20mpg)
Transmission:	Four-speed manual
Brakes:	Four-wheel Dunlop discs
Body/chassis:	Centre monocoque with separate front subframe

UNITED KINGDOM

189

JAGUAR E-TYPE SERIES III

Consistently regarded as one of the most beautiful cars ever created, the E-type was Jaguar's finest hour, a sporting legend that will never be eclipsed.

The E-type was designed by Jaguar's Malcolm Sayer, although it carried over cues from the previous D-type and XKSS cars.

The E-type started life with Jaguar's much-loved six-cylinder XK engine, first seen in 1948. However, in order to keep it competitive for the 1970s, a new V12 was developed, and fitted in 1971. It was the first road Jaguar to use V12 power. Although the fuel was delivered by conventional carburettors – albeit four of them – ignition was electronic.

As well as this 2+2 hatchback coupe, the E-type came as a very desirable convertible two-seater Roadster as well. Previous to the Series III, there was also a two-seater coupe.

All E-types featured four-wheel disc brakes, but on the Series III, they were vented at the front. Power assistance was fitted as well.

This Series III car – last of the line – featured a number of detail differences from the original design. The headlamps were exposed instead of being placed under faired-in clear covers, indicators were bigger, there was a larger air intake to help cool the V12 engine and the wheel arches were flared.

Even in the field of supercars, few cars are so respected or loved like the Jaguar E-type. It constantly tops the charts as the most beautiful car ever made, and on many people's personal wish lists, it remains number one. The E-type is, purely and simply, one of the greatest motoring icons of all time.

The 1960s was the decade in which British culture and style seemed to envelop the world, and the E-type was symbolic of this fantastic era. Launched in 1961 in

For the Series 3 E-type, a V12 engine was fitted. Despite the extra six cylinders, the lightweight aluminium construction of the V12 meant that it was still lighter than the old XK unit.

Geneva, its appeal was immediate and impressive. The 241km/h (150mph) performance was stunning, yet the purchase price was well below the usual cost for this level of automotive excellence.

Developments followed thick and fast. For 1964, there was the bigger 4235cc (258ci) engine, and, thanks to a longer wheelbase, 1966 saw the advent of a 2+2 E-type. One of the Jaguar's big markets was America, and so when federal regulations dictated it, the Series II had to adopt raised, exposed headlamps and larger bumpers from 1968.

DOUBLE THE CYLINDERS

Jaguar spent two million pounds developing a V12 engine for the E-type, unveiling it in the Series III of 1971. Although less pretty than the 1961 original, the Series III was more in-keeping with the 1970s, brimming with machismo and bulging with power. The V12 restored the E-type as one of the leaders of Europe's supercar monarchy.

All great things come to an end, though, and the E-type passed into history in 1975. It may be gone, but it will never be forgotten.

Jaguar E-type Series III

Top speed:	241km/h (150mph)
0–96km/h (0–60mph):	6.8 secs
Engine type:	V12
Displacement:	5343cc (326ci)
Max power:	186kW (250bhp) @ 6000rpm
Max torque:	390Nm (288lb-ft) @ 3500rpm
Weight:	1541kg (3390lbs)
Economy:	5.31km/l (15mpg)
Transmission:	Four-speed manual or three-speed automatic
Brakes:	Four-wheel discs, vented at front, solid at rear
Body/chassis:	Unitary monocoque construction with steel two-door coupe or roadster body

UNITED KINGDOM

193

JAGUAR XJ220

Unveiled to tremendous acclaim at the 1988 British Motor Show, the XJ220 was proof that Jaguar could still build jaw-dropping cars if it really tried.

The engine was based on that of the humble Austin Metro family hatchback. That said, the Metro in question was the 6R4 V6 competition car, far removed from the ordinary road car. It was fitted with twin Garrett T3 turbochargers, and the maximum power from the mid-mounted 24-valve unit installed in the XJ220 was 404kW (542bhp).

Bodies were hand-built out of lightweight – and unfortunately easily damaged – aluminium.

The big rear wing created a huge downforce at high speed, aided by two ground-effect venturis on the underside of the car.

It may have been a supercar, but the XJ220 did not skimp on the creature comforts inside. There was leather upholstery and carpeting throughout, plus a very effective sound system for owners not satisfied with the soundtrack from the exhausts.

The United States did not experience the XJ220. The car could not be engineered to meet stricter federal regulations, so Jaguar's biggest marketplace missed out on the fastest Jag ever built.

An XJ220 was certainly a car for show-offs. The engine cover was made of glass, so the V6 underneath could be admired!

UNITED KINGDOM

195

One of the most exciting British supercars ever built started out as a sketch on the back of a Christmas card. And without a small group of committed Jaguar engineers working in their own time, it would probably never have gone into production. The story behind the Jaguar XJ220 is one of dedication, but ultimately failure.

It was Jim Randle, the former head of Jaguar's product development department, who conceived the idea of a British sports

For a V6 engine, the XJ220's compact unit put out an almost unfeasibly high power output. How does 404kW (542bhp) sound from just a 3491cc (213ci) grab you?

car to challenge the world. His vision was shared by other company engineers, who built a prototype in their own time.

The result was sensationally unveiled at the 1988 British Motor Show, and to say it stole the show would be an understatement. Once everybody saw the amazing new Jaguar, nobody bothered to look at the new Ferrari F40 on a stand directly opposite it.

FASTER THAN FERRARI

Ferrari had even more cause to be worried when it was announced the XJ220 would go into production, albeit with cost-saving measures implemented. The original V12 engine was replaced by a V6, although it was still more powerful than the motor it replaced. Finally, in 1992, deliveries of the fastest road car in the world began to a privileged few.

Jaguar's car was superb, its timing less so. At $706,000, the XJ220 was never going to be a big seller. But the world economy was in recession when it appeared, and customers pulled out. Production was intended to reach 350, but just 281 were made before a struggling Jaguar admitted defeat in 1994.

Jaguar XJ220

Top speed:	335km/h (208mph)
0–96km/h (0–60mph):	3.8 secs
Engine type:	V6
Displacement:	3491cc (213ci)
Max power:	404kW (542bhp) @ 7200rpm
Max torque:	643Nm (475lb-ft) @ 4500rpm
Weight:	1473kg (3241lbs)
Economy:	4.57km/l (12.9mpg)
Transmission:	Five-speed manual
Brakes:	Vented discs, front and rear, with four-piston callipers
Body/chassis:	Aluminium alloy honeycomb monocoque with alloy two-door, two-seat body

UNITED KINGDOM

JAGUAR XKR

At last, the great XK name was back! The XKR continued where the legendary E-type left off, and restored Jaguar to the forefront of sports car manufacturers.

The convertible features a power-operated soft top, capable of folding away or erecting in just 20 seconds.

The V8 is all-alloy, with 32-valves and four camshafts, plus variable timing. More power is provided by the mechanically driven Eaton M112 supercharger.

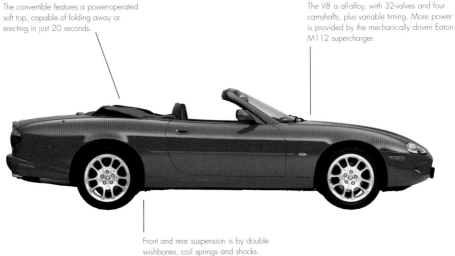

Front and rear suspension is by double wishbones, coil springs and shocks. Traction control also comes as standard.

An XKR convertible starred in the memorable ice chase sequence of the 2002 James Bond film *Die Another Day*, driven by villain Zao. Despite being fitted with special snow tyres, a Gatling gun, front and side-mounted missiles and mortar bombs in the boot, it was no match for 007's Aston Martin Vanquish. This offensive weaponry does not come as standard on most XKRs!

Despite its sporting credentials, the XKR does not stint on luxury. It has leather and wood inside, as well as air-conditioning, cruise control and an advanced sound system, among other toys.

There are echoes of the E-type's frontal treatment in the XKR's oval grille. The model is distinguished from the XK8 by having chrome mesh over its main air intake.

UNITED KINGDOM

When the iconic E-type stopped being produced in 1975, its replacement was the XJS. 'Replacement', though, is probably not quite the right word. Although successful in its own right, the XJS was more grand tourer than sports car, and never generated quite the same level of fanaticism and adulation as its illustrious predecessor.

The takeover of Jaguar by Ford in 1989 meant that there was finally enough money to develop a true and worthy successor to

The all-alloy V8 of the XKR is surprisingly lightweight. The 'Supercharged' on top of the camshafts denotes the fitment of an Eaton M112 supercharger, which increases power by almost a quarter.

the E-type. And when, in 1996, Jaguar launched its new XK8 sports coupe and convertible at the Geneva Motor Show (the same venue where it had first shown the E-type, 35 years before), most enthusiasts agreed it had achieved this aim.

At the heart of the new car was a technologically advanced V8 engine, clothed in a sleek, curvaceous body that captured the hereditary Jaguar look well. There was a big dose of luxury inside as well. In all ways, the XK8 was what a sporting Jaguar should be.

PUTTING THE 'R' IN

In 1998, the XK8 was taken to a new level, with the unveiling of the XKR. This more extreme version featured new suspension, bigger wheels and tyres, and an improved transmission. But the headline news was the Eaton supercharger now bolted to the V8 engine, which boosted power by 28 per cent. It was an exciting and very fast development. Top speed was electronically limited to 249.5km/h (155mph), but untethered the XKR would have been capable of so much more.

Interest in the Jaguar remains high as production continues into the 21st century.

Jaguar XKR

Top speed:	249.5km/h (155mph)
0–96km/h (0–60mph):	5.1 secs
Engine type:	V8
Displacement:	3996cc (244ci)
Max power:	276kW (370bhp) @ 6150rpm
Max torque:	524Nm (387lb-ft) @ 3600rpm
Weight:	1750kg (3850lbs)
Economy:	4.96km/l (14mpg)
Transmission:	Five-speed automatic
Brakes:	Four-wheel vented discs
Body/chassis:	Unitary monocoque construction with steel coupe or convertible body

UNITED KINGDOM

LAMBORGHINI MIURA

Not only did it look spectacular, but the epic Miura was also the first supercar with a mid-mounted engine, and a quad-cam V12 at that.

A stylish touch was the door handles blending into the cooling slats positioned into the door frame.

Visibility through the wide, raked windscreen was extremely good. Trying to see where you had come from, though, was less easy. The engine cover had slats over it, used for extracting hot air from the engine, which made the rear view very poor.

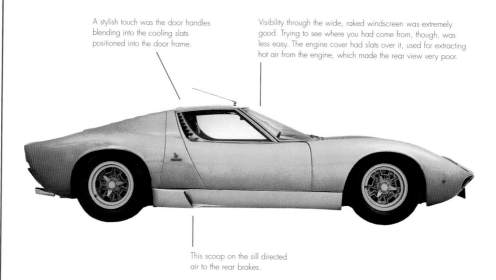

This scoop on the sill directed air to the rear brakes.

A trademark feature of the Miura were its 'eyebrows' surrounding the folding headlamps. They looked like cooling vents, but were purely cosmetic. They were deleted on later cars.

The V12 was transversely mounted immediately behind the passenger cabin. This arrangement was actually inspired by the Mini, which also had a transverse engine, albeit one of slightly less power! The transmission was positioned behind the power unit and (also as on the Mini) shared the same oil as the engine.

Because of its immensely strong chassis, only the central tub of the Miura was made of steel. The bonnet and engine cover were made of alloy, and both were flip-up panels.

ITALY

203

nzo Ferrari said: 'You don't have the slightest idea of how to drive a Ferrari. You'd rather drive your tractors.' He probably came to regret this outburst. The object of his scorn – tractor manufacturer Ferruccio Lamborghini – swore revenge, and established a rival marque to get it.

Just three years after that encounter, Lamborghini built the awesome Miura, the world's first mid-engined supercar. It was a creation that did not so much put Ferrari in the shade as totally eclipse it. Revenge has rarely been so dramatic…or so radically-shaped and fast.

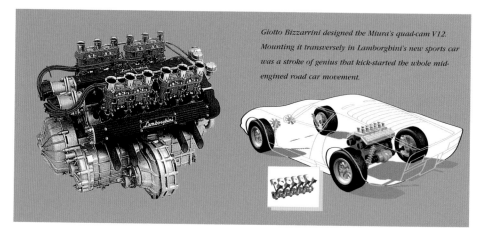

Giotto Bizzarrini designed the Miura's quad-cam V12. Mounting it transversely in Lamborghini's new sports car was a stroke of genius that kick-started the whole mid-engined road car movement.

Proving a point, Lamborghini sought out ex-Ferrari employees to turn his vengeful dreams into reality, including Giotto Bizzarrini, who had been responsible for the Ferrari 250 GTO. It was Bizzarrini who came up with the quad-cam V12 engine that would be used by Lamborghini for years to come.

FERRARI BEATER

Meanwhile, Marcello Gandini of Bertone designed the elegant, flowing Miura body, an automotive sculpture of grace, power and style. The mid-engine layout endowed it with new levels of handling and poise, and perhaps more importantly for Lamborghini, Ferrari had absolutely nothing to answer it. *Car* magazine called it 'far and away the most exciting production development since the war, an inspired creation, which will undoubtably become a classic, fit to stand beside the most delectable possessions man has yet succeeded in manufacturing for his delectation'. Praise indeed.

The Miura, also known as the TP400, lasted until 1973, with uprated versions being launched in 1970 (the TP400S) and 1971 (the TP400SV). Its successor was the equally remarkable Countach.

Lamborghini Miura

Top speed:	277km/h (172mph)
0–96km/h (0–60mph):	6.9 secs
Engine type:	V12
Displacement:	3929cc (240ci)
Max power:	496kW (370bhp) @ 7700rpm
Max torque:	388Nm (286lb-ft) @ 5500rpm
Weight:	1296kg (2851lbs)
Economy:	4km/l (11.2mpg)
Transmission:	Five-speed manual
Brakes:	Four-wheel solid discs
Body/chassis:	Steel monocoque platform with steel and alloy two-door, two-seat coupe body

ITALY

LAMBORGHINI ESPADA

One of the most unusually shaped supercars ever built, the four-seater, hatchback Lamborghini Espada was a striking blend of surprising practicality and exhilarating performance.

Unlike most 'two-plus-two' supercars, the Espada was a genuine four-seater, and anyone sitting in the back could enjoy a reasonable degree of comfort.

The slots down the side of the hood were crucial features to try and keep the monstrous V12 engine cool, venting hot air from the engine bay.

Lamborghini turned to Marcello Gandini of Bertone to style the Espada. His first design for the marque had been the beautiful Miura, so Lamborghini obviously believed he had proved his credentials. The Espada was actually based on a show car that had appeared in 1967 at the Geneva Motor Show.

Power came from a development of the 24-valve V12 quad-cam engine used in previous Lamborghinis, with six Weber carburettors used to feed fuel. Balancing all of those properly was a very tricky task.

The low, sloping roof line meant an ordinary boot could not be fitted, so a big glass hatchback was adopted instead. Its usefulness was compromised by the shallow load space, though. Lamborghini even made special luggage sets to fit the limited area.

ITALY

L amborghini continued its bitter rivalry with Ferrari by launching the Espada in 1968. Just two years after the Miura, with its mid-mounted engine, had so surprised its competitor, the Raging Bull kept the Prancing Horse on its toes by announcing the four-seater Espada.

Ferrari simply had nothing like this 241km/h (150mph) executive express in its armoury. In fact, for sheer brazenness and idiosyncratic looks, no car manufacturer had anything remotely resembling the Espada.

In 1967, Bertone displayed a car called the Marzal at the Geneva Motor Show. Interest in

In the Espada, twin distributors were fitted to Lamborghini's hereditary V12 engine, along with six Weber carburettors, four camshafts and a total of 24 valves.

it was enough to convince Lamborghini to rework it into the Espada the following year, replacing the six-cylinder engine with a traditional V12 in the process. To accommodate the rear seats, the car's wheelbase was 10cm (4in) longer than previous front-engined Lamborghinis.

Not everybody was impressed by the distinctive looks, which could look ungainly from some angles. But, love it or loathe it, the Espada was an eye opener, turning heads wherever it appeared. As the most expensive Lamborghini available at the time, it certainly needed to.

1970S CHANGES

Lamborghini bought out an improved Series 2 version (the 400GTE) in 1970, with brake, drive and power enhancements, and then followed it up after another two years with the Series 3, now boasting 272kW (365bhp). An automatic transmission, courtesy of the Chrysler three-speed Torqueflite gearbox,

became available at the same time, and the much-needed power steering option was standardized.

The Espada was built for another five years, becoming one of the marque's best-sellers, with 1217 examples built.

Lamborghini Espada

Top speed:	241km/h (150mph)
0–96km/h (0–60mph):	8.0 secs
Engine type:	V12
Displacement:	3929cc (240ci)
Max power:	261kW (350bhp) @ 7500rpm
Max torque:	393Nm (290lb-ft) @ 5500rpm
Weight:	1762kg (3876lbs)
Economy:	3.33km/l (9.4mpg)
Transmission:	Five-speed manual
Brakes:	Vented discs, front and rear
Body/chassis:	Steel monocoque with alloy hood. Two-door, four-seat hatchback body.

ITALY

LAMBORGHINI COUNTACH

There are few supercars as utterly dramatic as the Countach. Just standing still, it is a sensational looking creation. In action, it is awe-inspiring.

This unusual trapezoidal rear wheel arch shape lasted until 1978. When the LP400S Countach came out in 1978, a more conventional shape was adopted, albeit flared to accommodate fatter tyres.

Compared to later Countachs, the lines of this original LP400 model were simplistic and pure, without the spoilers and body kit extensions that would go on to clutter later cars.

There were no obvious door handles. They were actually hidden inside the air ducts on the side of the car, to avoid interrupting the slab-sided lines of the body. The doors, which are made of fibreglass, had to swing upwards when opened, because the sills were so large.

Positioned immediately behind the driver was the mighty V12 engine of 3929cc (240ci), with the fuel fed by a bank of six Weber carburettors. It was difficult to keep cool, so the Countach had two radiators mounted on either side of the engine.

The Countach had two boots, one in front, and one behind the engine in the rear. However, neither had much capacity, thanks to the uncompromisingly wedge-shaped body.

ITALY

211

It is, without doubt, one of the most awe-inspiring cars ever created. From any angle, the Lamborghini Countach is a brutally imposing vehicle, one that is not so much built as chiselled out of aluminium and fibreglass and brought to life with a V12 heart, like some kind of automotive Frankenstein's monster.

Nothing like the Countach had been seen before. For that matter, nothing much like it has been seen since. Car design at the start of the 1970s was moving towards wedge-

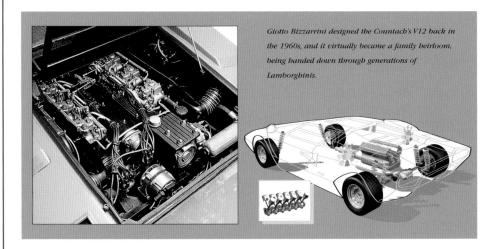

Giotto Bizzarrini designed the Countach's V12 back in the 1960s, and it virtually became a family heirloom, being handed down through generations of Lamborghinis.

shaped styling, but with the razor-edged Countach, designer Marcello Gandini of Bertone took a giant leap into the abyss. It was a gamble that paid off, the Lamborghini's radical looks earning it a legion of fans – most of whom could never afford it.

PROTOTYPE TO SUPERCAR

It was a 1968 Bertone show car, known as the Carabo, that inspired the Countach. Three years later, at the Geneva Show, the Countach prototype was unveiled in all its angular glory. The V12-powered production version took a further three years to arrive, after Lamborghini decided to use a tubular spaceframe chassis instead of the monocoque originally intended.

The first LP400 Countachs were somewhat plain and unadorned, but in 1978 came the LP400S, sporting flared wheel arches and fender extrusions to give an even more menacing look. Four years later, and the LP500 upped power to 4800cc (293ci).

The final evolution of the Countach was the 5000 QV *Quattrovalvole* of 1985, with an even bigger engine, and updated looks, although it was still resolutely Lamborghini in style. The Diablo superseded it in 1991.

Lamborghini Countach

Top speed:	290km/h (180mph)
0–96km/h (0–60mph):	5.5 secs
Engine type:	V12
Displacement:	3929cc (240ci)
Max power:	280kW (375bhp) @ 8000rpm
Max torque:	360Nm (266lb-ft) @ 5000rpm
Weight:	1373kg (3020lbs)
Economy:	4.96km/l (14mpg)
Transmission:	Five-speed manual
Brakes:	Vented discs, front and rear
Body/chassis:	Separate tubular-steel chassis with two-door, two-seat alloy and fibreglass body

ITALY

LAMBORGHINI DIABLO

The name means 'Devil' in Italian, a very applicable title for one of the most menacing-looking cars ever launched, a worthy successor to the Countach.

This panel is actually the removable roof, which could be secured in this position above the engine cover when not in place.

The Diablo continued the low-nose, high-backed wedge theme of the previous Countach. Also carried over from the previous Lamborghini supercar were the doors, which opened upwards.

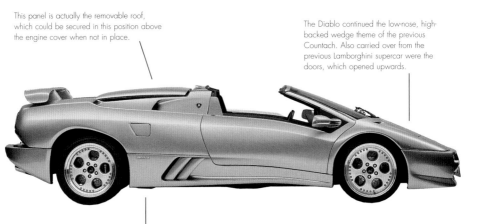

Stylist of the Diablo was Marcello Gandini, who had also designed the Miura and Countach for Lamborghini, so he had a proven track record when it came to designing cars for the Raging Bull marque.

Beneath the bodywork was a multi-tube chassis, similar to that used on the Countach. That car had round tubing, though, while on the Diablo, they were square.

It is a Lamborghini, so it must be a V12! The quad-cam all-alloy, immensely mighty power unit dated back to 1963, but was much-developed by the time it ended up in the Diablo. Among the modifications were four valves per cylinder, making a total of 48.

Diablos were available in both rear-wheel and four-wheel drive versions. This example is all-wheel drive, as denoted by the VT badging on the rear.

It cannot be easy for a company like Lamborghini to launch a new car. Because of the marque's tradition for creating concoctions that are extraordinary in looks and performance, each new Lamborghini has a lot to live up to. Just how do you replace such a monumental icon like the Countach?

The answer was the Diablo, the supercar that took Lamborghini into the 1990s with extreme style. Perpetuating the established mid-engined, wedge-shaped Lamborghini

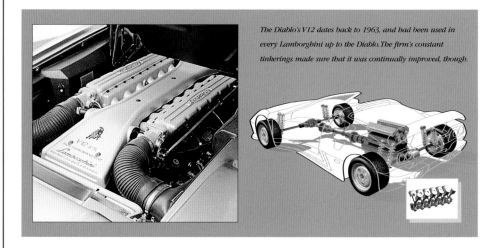

The Diablo's V12 dates back to 1963, and had been used in every Lamborghini up to the Diablo. The firm's constant tinkerings made sure that it was continually improved, though.

tradition, designer Marcello Gandini created a devilishly handsome beast that fully justified its Diablo tag. It was heaven in hellish form. Had it not been for Chrysler buying Lamborghini in 1987, the Diablo would have been just a reworked Countach. The big conglomerate brought financial security plus technological expertise, meaning that 1990's Diablo was a completely new car. Lamborghini launched it to considerable fanfare in Monte Carlo: in terms of power and sheer presence, it was over the top. Once again, Ferrari was left floundering in its rival's exhaust fumes.

EXCITING DEVELOPMENTS

For 1992, Lamborghini took the top off, to create the Diablo Roadster, although it would be three more years before the cars were actually put into production. Nothing without a fixed roof was faster on the roads. The Roadster featured permanent four-wheel drive, as introduced on the Coupe in 1993, although by 1998, a two-wheeled open top version was available too.

The Diablo was dropped in 2001 for the Murcielago.

Lamborghini Diablo

Top speed:	325km/h (202mph)
0–96km/h (0–60mph):	5.1 secs
Engine type:	V12
Displacement:	5707cc (348ci)
Max power:	395kW (530bhp) @ 7100rpm
Max torque:	604Nm (446lb-ft) @ 5500rpm
Weight:	1629kg (3583lbs)
Economy:	4.07km/l (11.5mpg)
Transmission:	Five-speed manual, permanent four-wheel drive or rear-wheel drive
Brakes:	Four-wheel vented discs
Body/chassis:	Tubular-steel chassis with alloy, composite and carbon fibre two-door body

ITALY

LANCIA STRATOS

The styling is uncompromising, so is the driving experience. The Stratos is a car that will turn and bite, unless a driver treats it with respect.

The spoiler served three purposes. Primarily, it generated downforce to make the car more stable at high speeds, but it also helped direct cooling air into the engine bay. In the event of an accident, it also acted as a rollover bar.

The body design of the Stratos was the responsibility of the Italian Bertone coachbuilding house. It was one of the most extreme shapes ever to appear on a road car, having started out as a show concept at the start of the 1970s.

The Stratos had a folded sheet-steel frame, with a fibreglass body built around it. This lightweight material could be used since none of the panels were load-bearing. The front and rear exterior sections, held down by straps, could be completely lifted off.

It was a tight squeeze inside the cockpit, and the offset pedals did not make driving any easier. You had to be dedicated to pilot a Stratos.

Power for the Stratos came from a quad-cam Ferrari Dino engine, Lancia having access to the compact but powerful V6 since it was also owned by Ferrari's guardian, Fiat.

ITALY

219

The Stratos is one of those cars which is appreciated far more now than when it was in production. Less than 500 were built between 1973 and 1975, and Lancia was still trying to sell it until 1980. It was a noisy, uncomfortable and incredibly difficult car to drive, especially at speed. Aesthetically, it could never be regarded as conventionally beautiful. Worse, it took a superb driver to get the best out of a Stratos – and there were few who seemed equal to the challenge.

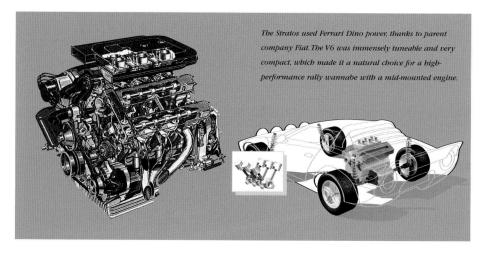

The Stratos used Ferrari Dino power, thanks to parent company Fiat. The V6 was immensely tuneable and very compact, which made it a natural choice for a high-performance rally wannabe with a mid-mounted engine.

Why, then, is the Stratos such a legend now? Well, when it came to rallying in the 1970s, there was absolutely nothing to touch the Stratos. It dominated rallying like nothing else, winning the World Championship in 1974, 1975 and 1976. On any surface, it was the master…assuming its driver was skilled enough to handle it.

BERTONE WEDGE

The low-slung, dramatically wedge-shaped creation first saw the light of day in 1970 as a Bertone concept. It was refined for the 1971 show circuit, creating such interest and showing such potential that Lancia decided it would be a great basis for a rally car.

In order to meet homologation rules, 500 road cars also had to be built. Whether the full quota was ever built is debatable, as it was affected badly by the fuel crisis of the 1970s, which harmed sales. It may have been much admired on the track, but it took far longer to be esteemed off it.

Ironically, its character and idiosyncrasies are what now make it one of the most desirable supercar classics today.

Lancia Stratos

Top speed:	225km/h (140mph)
0–96km/h (0–60mph):	7.0 secs
Engine type:	V6
Displacement:	2418cc (148ci)
Max power:	142kW (190bhp) @ 7000rpm
Max torque:	225Nm (166lb-ft) @ 5500rpm
Weight:	982kg (2161lbs)
Economy:	6km/l (17mpg)
Transmission:	Five-speed manual
Brakes:	Four-wheel vented discs
Body/chassis:	Fibreglass two-door, two-seat coupe with folded sheet-steel frame

ITALY

221

LISTER STORM

Lister started with a Jaguar V12 engine, and built a supercar around it. The very fast, superbly handling result was then aptly christened the Storm.

The composite monocoque construction is very advanced. It is an alloy honeycomb, with five bulkheads incorporated. The body panels are made of carbon fibre, and rivetted to the honeycomb.

Lister has had a long association with Jaguar, dating back to the 1950s. Its Storm uses the marque's V12 engine, but with even extra muscle. Practically everything that can be altered is modified, with the displacement increased from 5994cc (366ci) to a whopping 6997cc (427ci). A belt-driven supercharger is also added to each cylinder bank.

The sophisticated suspension is double wishbones at the front, with a multi-link rear, plus anti-roll bars at both ends. Camber and caster can be adjusted.

222

Although it is front-engined, the Storm's V12 is mounted as far back
as possible, with most of it behind the front wheels. That makes
balance almost perfect. The power reaches the rear wheels via a
Getrag six-speed manual gearbox.

The underneath of the Lister, with its central spine,
is designed to generate downforce to keep the
car stuck to the road at speed.

223

The founder of Lister, Brian Lister, had worked with Bristol, Chevrolet and Maserati engines before his successes with Jaguar. His series of Jag-engined racing sports cars during the 1950s were hugely successful, but production ended in 1959.

However, the Lister company was back again in the 1980s. Specially commissioned by Jaguar, it started tuning XJS engines for racing. This was the first step down the road that would eventually lead to the outfit building the Storm.

The Storm's V12 engine is based on a standard Jaguar item. However, Lister completely rebuilt the unit, increasing its cubic capacity thanks to a far longer stroke. Only the single-camshaft head stayed the same.

The Jaguar work was lucrative, and by 1991, the company had enough money in the bank to think seriously about building its own car. Its approach was novel. Rather than styling the car, and then fitting the mechanical components inside, the Storm was built around the optimum position for the Jaguar V12 engine. As Lister Cars boss Laurence Pearce explained, 'Having determined these vital points, it was essentially a question of seeing where the rest would go.'

BLISTERING LISTER

Considering its engineering-led principles, the Storm of 1993 was surprisingly practical. It may not have been the most attractive shape around, but it had a definite presence and its drag co-efficient was a low 0.32. It could also seat four occupants along with their luggage. More importantly for a supercar, though, it was capable of preposterous speeds, and with the driver's foot flat to the floor, 96km/h (60mph) came up in just 4.1 seconds.

The road cars are still built, the GT racing cars still successful. The Storm won the FIA GT World Championship in 2000.

Lister Storm

Top speed:	322km/h (200mph)
0–96km/h (0–60mph):	4.1 secs
Engine type:	V12
Displacement:	6996cc (427ci)
Max power:	443kW (594bhp) @ 6100rpm
Max torque:	786Nm (580lb-ft) @ 3450rpm
Weight:	1440kg (3169lbs)
Economy:	4.25km/l (12mpg)
Transmission:	Six-speed manual
Brakes:	Four-wheel vented discs
Body/chassis:	Composite construction comprising alloy honeycomb monocoque with carbonfibre roof and body panels

UNITED KINGDOM

LOTUS ESPRIT V8

The Esprit's final development was arguably the best, a V8 engine giving it the performance and power it had screamed out for from the beginning.

The original straight-edged wedge shape of the Esprit was designed by Giorgetto Giugiaro back in the 1970s. For 1987, the body was restyled by Peter Stevens. Its new appearance was softer, more curvaceous, yet at no point was the new profile more than 25mm (1in) away from the old shape.

Squeezing in the mid-mounted V8 engine into the tight engine bay meant the radiator had to be installed in the front boot, steeply angled to fit under the raked nose.

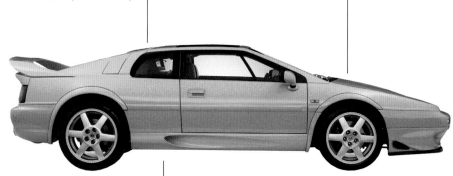

Bodies were built of fibreglass, mounted on Lotus' traditional steel backbone chassis. The Esprit also had tough Kevlar reinforcement around the roof pillars and the sills, to make the structure stiffer and also give increased protection in the event of an accident.

After 20 years of being fitted with a four-cylinder engine, the Esprit got Lotus' first ever production V8 in 1996. The powerful, all-alloy unit was also used in racing. Continuing the previous Esprit series' turbocharged theme were two small Garrett T25 turbos.

As the rear tyres did most of the work, they were wider than the front ones.

At the beginning of 2003, Lotus afficionados went into mourning when it was announced that one of the most enduring supercars of all time was going out of production. After well over a quarter of a century, the Esprit was to be retired, overshadowed by the newer Elise and Exige. This remarkable car had survived through four decades, and although the last models were very different from the first, it is remarkable just how visually similar a 1976 Esprit is compared to one from 2003.

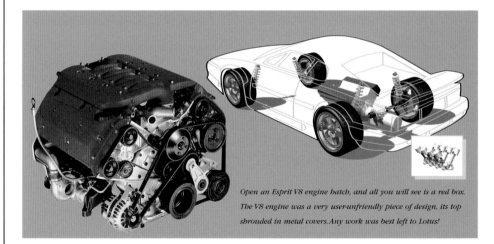

Open an Esprit V8 engine hatch, and all you will see is a red box. The V8 engine was a very user-unfriendly piece of design, its top shrouded in metal covers. Any work was best left to Lotus!

There is simply no way of mistaking the family resemblance.

The Giugiaro-styled Esprit got off to a good start. A year after its launch, it was driven – both on road and underwater – by James Bond in *The Spy Who Loved Me*. Publicity like that ensured tremendous interest in the car, but the Esprit was a tremendous package in its own right, its looks and superb handling gaining it many plaudits.

FOUR-CYLINDER TO V8

However, reliability at first was poor, and the small 1973cc (120ci) mid-mounted engine of just four cylinders proved to be a disappointment. In the years that followed, engine size was upped and turbocharging introduced, and the Esprit gradually became a more dependable car.

In 1987, it was subtly restyled to keep it fresh, but the biggest news of all was the V8 engine transplant of 1996. Developed for racing, this all-new plant was detuned for installation in the Esprit, but even so, it brought a massive increase in power and flexibility, thereby giving the Esprit – at last – the kind of Ferrari-beating potential it had long deserved.

Lotus Esprit V8

Top speed:	277km/h (172mph)
0–96km/h (0–60mph):	4.2 secs
Engine type:	V8
Displacement:	3506cc (214ci)
Max power:	260kW (349bhp) @ 6500rpm
Max torque:	400Nm (295lb-ft) @ 4250rpm
Weight:	1349kg (2968lbs)
Economy:	6.94km/l (19.6mpg)
Transmission:	Five-speed manual
Brakes:	Four-wheel vented discs, ABS
Body/chassis:	Sheet steel fabricated backbone chassis with fibreglass two-door coupe body

UNITED KINGDOM

MASERATI GHIBLI

Maserati never quite managed to scale the same heights as Ferrari and Lamborghini, but its gorgeous Ghibli was almost the equal of its Italian rivals.

Maserati fitted two fuel tanks, which could be filled via flaps on either side of the roof pillars.

Under the hood lurked Maserati's family quad-cam V8, first seen in the marque's 450S racing cars of the 1950s. It was detuned from competition standard for installation in the Ghibli, but still produced a heady 276kW (370bhp).

There was no attempt to save weight on the Ghibli, Maserati's view being that the Ghibli's engine was more than powerful enough. Thus the body was made completely out of steel, instead of using a lighter material. The unusual-looking wheels were made of alloy, though.

Pop-up headlamps were a novelty when the Ghibli appeared, but practically all supercar manufacturers would soon be using them.

As well as the coupe, a very limited number of convertibles were built. Fixed-head versions outnumbered these more desirable Spider versions by almost 10 to one.

The sleek styling was by Italian master Giorgetti Giugiaro, then just a young designer at coachbuilder Ghia. It was one of the designs he was most proud of, and helped bring him to greater prominence.

ITALY

231

It was financial and competition misfortune that turned Maserati from a racing car maker into a supercar manufacturer. The team had been active in competition from the 1920s, but in the 1950s, the firm was badly affected by some serious racing accidents, as well as monetary difficulties. So it decided to withdraw from racing, and concentrate on building high-performance road cars instead.

Without doubt, the greatest and best-loved of all these subsequent Maseratis was the Ghibli of 1966. Although it lacked the V12 technical excellence of its contemporaries

Maserati had a strong racing heritage, and used a detuned version of its 1950s quad-cam V8 alloy racing engine in the Ghibli.

from Ferrari and Lamborghini, its breathtaking looks and extreme level of V8 power, coupled with its luxury appointments, made it a sales winner. More Ghiblis managed to find homes than its immediate rivals, the Ferrari Daytona and Lamborghini Miura.

BASIC BUT BEAUTIFUL

Under the Giugiaro-designed steel skin, with its radically low, shark-shaped nose, the Ghibli was largely conventional, a quad-cam engine mounted at the front to drive the rear wheels. Although the suspension was well-behaved wishbones at the front, the rear kept an old-fashioned live axle and leaf spring arrangement, meaning the Ghibli could be a handful on fast corners. In most other respects, though, it was an easier machine to get to grips with than a Ferrari or Lamborghini.

The debut for the Ghibli was at the prestigious Turin Motor Show in 1966, where it grabbed all the headlines. The coupe went into production a year later, followed in 1969 by the open top Spider. 1970's more flexible SS version upped strength and torque, before the Ghibli passed into history in 1973.

Maserati Ghibli

Top speed:	248km/h (154mph)
0–96km/h (0–60mph):	6.8 secs
Engine type:	V8
Displacement:	4719cc (288ci)
Max power:	276kW (370bhp) @ 5500rpm
Max torque:	441Nm (326lb-ft) @ 4000rpm
Weight:	1702kg (3745lbs))
Economy:	3.9km/l (11mpg)
Transmission:	Five-speed manual or three-speed automatic
Brakes:	Vented discs
Body/chassis:	Steel two-door 2+2 coupe or convertible with tubular-steel chassis

ITALY

233

MASERATI BORA

Maserati's first mid-engined effort was the Bora, which managed to be fast, quirky and individualistic, thanks to the involvement of Citroën and Giorgetto Giugiaro.

Giorgetto Giugiaro, who had designed the Ghibli, was brought back to style the Bora and its smaller sister, the Merak. The flying buttresses at the rear of both cars were notable features.

Pre-production Ghiblis had all-alloy bodies, but when the standard car started to be manufactured, steel was used throughout. The front two-thirds of the Bora were of monocoque construction, with the rear a tubular steel structure to support the engine, transmission and suspension.

In order to make maintenance on the confined V8 easier, the entire rear of the car could be completely removed.

Maserati's versatile quad-cam alloy V8, designed in the 1950s, made yet another appearance in the Bora. This time, though, it was mid-mounted (with the ZF transmission behind it) for the first time, following the trend of other supercar manufacturers. With its hemispherical combustion chambers and low-revving qualities,it was more American than European in nature.

Citroën hydraulics featured in the Bora, and although their use did not extend to the suspension, the brakes and power-operated pedals, seats, headlamps and steering column were all controlled by hydro-pneumatics.

ITALY

If it had not been for Citroën, Maserati's Bora might never have been built and one of the most unconventional mid-engined supercars would never have appeared. Lamborghini had taken the mid-engine lead with its Miura of 1966, and Ferrari had followed suit with the Dino of 1967. Maserati was expected to come up with the goods next: with the V8 Bora, it did not disappoint, even adding a little bit extra.

Citroën's 1968 takeover of Maserati brought a welcome injection of resources.

The Bora's V8 engine is a low-revving unit, generating its maximum torque low down in the rev range. Its installation in the Bora was the first time Maserati had experimented with mid-mounted engines.

Out of this new regime came the Bora (named, in Maserati fashion, after a strong wind). Unveiled in Geneva in March 1971, it attracted attention thanks to its singular design from Giugiaro's new Ital Design agency: wedge-shaped, it had a large glass area punctuated by flying buttresses enclosing the rear. Inside, Citroën eccentricities abounded, thanks to the use of hydraulics to control the sharp brakes and lots of toys for the driver. It was a car with plenty of character, as well as a mean turn of speed and pinpoint handling.

EXTENDED FAMILY

The Merak, a smaller-engined and cheaper version with Citroën's SM V6 engine, joined its big sister in 1972. It looked very similar to the Bora, but without the extensive glasswork at the back. In 1974, the Bora finally made it to America, thanks to an enlarged 4930cc (301ci) V8 that finally met US regulations.

Maserati was bought by De Tomaso in 1975, but the marque struggled. There were no more changes to the Bora and it was dropped in 1980.

Maserati Bora

Top speed:	257.5km/h (160mph)
0–96km/h (0–60mph):	6.5 secs
Engine type:	V8
Displacement:	4719cc (288ci)
Max power:	231kW (310bhp) @ 6000rpm
Max torque:	440Nm (325lb-ft) @ 4200rpm
Weight:	1623kg (3570lbs)
Economy:	3.54km/l (10mpg)
Transmission:	Five-speed ZF manual
Brakes:	Vented discs with Citroën high-pressure hydraulics
Body/chassis:	Steel unitary construction front sections with square-tube rear frame, steel two-door coupe body

ITALY

237

McLaren F1

McLaren's extraordinary F1 could well be the ultimate road supercar of all time, built with few compromises by the famous world-beating Formula 1 team.

At the rear of the car was a retractable rear spoiler. However, unlike the Porsche 911, where the spoiler rises when a certain speed is reached, the F1's wing deployed when the brakes were applied heavily. This helped to stop the nose of the car from diving, and also generated downforce, bringing the tyres into closer contact with the road.

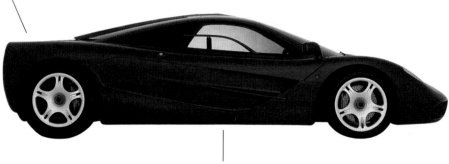

In an echo of Lamborghini (one of its main rivals), McLaren fitted the F1 with huge upward opening doors.

Inside, the driver was seated in the centre of the car, with two smaller passenger seats set further back on either side. A 386km/h (340mph) speedometer reminded the driver just what the F1 was capable of.

The central overhead spine sucked cooling air over the top of the car and directly into the engine.

Luggage space in the McLaren F1 was severely limited. However, behind each door, set into the exterior panels in front of the rear wheels, were two well-concealed hinged compartments which could take a little extra baggage, especially if you chose the option of McLaren's specially tailored luggage.

Imagine being in one of the world's fastest, most expensive supercars, and it breaks down on you? Not only would it be extremely embarrassing, but there would be the added problem of wondering exactly where you could get it fixed. And how long it would take...

Owners of the McLaren F1 did not, however, have that kind of worry. If their car broke down anywhere in the world, a McLaren mechanic would be on the next flight out to fix it. That was just one of the many features that marked the F1 out as something very, very special.

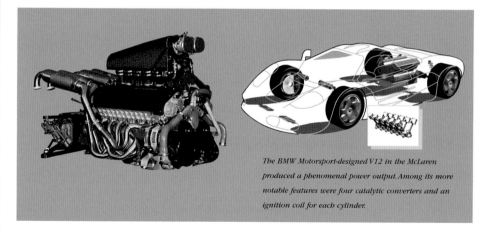

The BMW Motorsport-designed V12 in the McLaren produced a phenomenal power output. Among its more notable features were four catalytic converters and an ignition coil for each cylinder.

Just 100 F1s were built from 1993 to 1997, the first one being completed on Christmas Eve 1993. It must have been a very special present for one customer. Even 10 years after its first appearance, the F1 is still recognized as one of the most extraordinary supercars ever built, the ultimate no-limits performance car.

MURRAY'S VISION

Gordon Murray, McLaren's design chief, announced his scheme to construct his dream car in 1989. His position with one of the world's foremost Formula 1 teams meant that he was able to call on the talents of the very best. Peter Stevens, from Lotus, styled the car's sexy but menacing shape, while BMW Motorsport came up with a brand new V12 engine. The attention to detail throughout was exquisite.

Everything was complete by 1993, but with its million-dollar price tag, the F1 was an exclusive toy, and production failed to reach the intended 300 cars. This did not matter, though; the F1 had made its mark on automotive history.

McLaren F1

Top speed:	372km/h (231mph)
0–96km/h (0–60mph):	3.2 secs
Engine type:	V12
Displacement:	6064cc (370ci)
Max power:	467.5kW (627bhp) @ 7300rpm
Max torque:	649Nm (479lb-ft) @ 4000rpm
Weight:	1020.5kg (2245lbs)
Economy:	4.39km/l (12.4mpg)
Transmission:	Six-speed manual
Brakes:	Four-wheel Brembo discs
Body/chassis:	Carbon fibre two-door, three-seat coupe with carbonfibre and Nomex/alloy honeycomb monocoque chassis

UNITED KINGDOM

MERCEDES-BENZ 300SL

As the first true postwar supercar, the sensational gull-winged Mercedes-Benz of 1954 set the pattern for all the spectacular sports cars that would follow it.

Alloy 'gull-wing' doors – which were hinged at the top – had to be used on the 300SL due to its high sills. The height was necessary due to the novel spaceframe chassis, formed out of a network of small tubes giving the car great structural integrity but also low weight. This was the first time such a construction had been used on a road car. Indeed, until recently, it was not used again.

The door handles were unusual. At first glance, they are completely concealed by the bodywork, but press one end and they slide out! A nice touch.

The huge vents in the front fenders were a low-tech attempt to help keep the engine cool.

As well as the doors, the hood and boot lid were also made of alloy. The rest of the bodywork was steel.

One of the main innovations on the 300SL was its pioneering use of fuel injection. The mechanical Bosch system boosted the carburetted output of the six-cylinder engine by more than 100 per cent.

Drum brakes were used on the 300SL coupes; the later roadster boasted discs.

GERMANY

Whenever it had appeared, the Mercedes-Benz 300SL would have been recognized as a true giant among sports cars. But what made this street racer particularly amazing was that it appeared at the beginning of the 1950s, less than a decade after the end of the war that had left Mercedes-Benz practically in ruins.

Mercedes had a rich racing heritage, but only in 1952 did it build its first postwar racer, the 300SL coupe. Like its prewar counterparts, the 300SL was a consummate

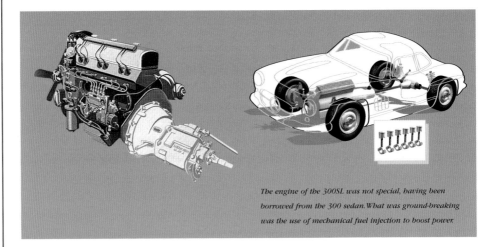

The engine of the 300SL was not special, having been borrowed from the 300 sedan. What was ground-breaking was the use of mechanical fuel injection to boost power.

professional on the race tracks, winning Le Mans on its first attempt.

FASTER THAN A RACE CAR

In February 1954, the magic came to the street. At the New York Motor Show, the 300SL was launched in road-going form, and this, unbelievably, was more dynamic than the racer. Carburettors were replaced by fuel injection, giving a huge boost in power. The mechanical system may have been primitive, but was a big step forward technically. The gull-wing doors grabbed headlines as well.

The 300SL was fabulously expensive, but buyers got a lot of car for their money. Thanks to its spaceframe chassis, the SL (it stands for 'super light') handled superbly and was capable of speeds approaching 266km/h (165mph).

After building 1400 coupes, Mercedes-Benz introduced the faster, but less exotic-looking roadster for 1957. The inefficient drum brakes were replaced in 1961 by disc brakes at last capable of reigning in all that power. Production ended in 1963, and today, the 300SL is one of the most desirable classic supercars around.

Mercedes-Benz 300SL

Top speed:	266km/h (165mph)
0–96km/h (0–60mph):	9.0 sec
Engine type:	In-line six
Displacement:	2996cc (183ci)
Max power:	179kW (240bhp) @ 6100rpm
Max torque:	293Nm (216lb-ft) @ 4800rpm
Weight:	1295kg (2850lbs)
Economy:	6.37km/l (18mpg)
Transmission:	Four-speed manual
Brakes:	Four-wheel drums, later four-wheel discs
Body/chassis:	Steel and alloy two-door coupe with steel spaceframe chassis

GERMANY

MG METRO 6R4

One of the most unlikely supercars ever built, the Metro 6R4 had the same shape as a small British hatchback, but its performance was utterly phenomenal.

The 6R4 was a 'silhouette' vehicle: something that looked like a standard production car, but was really radically different. This 'Metro' was made from carbonfibre and Kevlar, and featured a front air dam, side pods and dramatic roof spoiler. You certainly could not buy anything like it from your local Austin Rover dealership!

The side bodywork extrusions contained the two radiators. This saved space in the engine area, and made maintenance easier, but it also meant the radiators were vulnerable to damage during rallies.

Bizarrely, inside the 6R4 was the same dashboard and steering wheel as used in the ordinary MG Metro, although the interior was stripped out to save weight, and racing seats were fitted.

When the 6R4 was built in prototype form, it featured a Rover V8 with two cylinders lopped off to create a V6. Production versions had a specially designed quad-cam V6 with Lucas fuel injection. Unlike standard Metros, the engine was mid-mounted and drove all four wheels.

The whole of the rear panel could be removed if necessary, a common feature found on Group B rally cars.

Though small, the Austin Metro was a major car for British Leyland. Hopes were high that here was a product that could be an international success and lift BL out of the doldrums of the 1970s.

It was in this spirit of optimism that BL's Austin Rover division decided to re-enter motorsport, with a rally car based on the Metro. Bravely it chose to contest the Group B class, then one of the most fiercely competitive and downright ferocious forms of motorsport around.

As the Williams Formula 1 team was being sponsored by BL at the time, it got to build

The 6R4's V6 engine was completely removed from the standard Metro's four-cylinder A-series. In its ultimate tuned Evolution form, the V6 could output 306kW (410bhp).

the MG Metro 6R4, the task shared between chief engineer/designer Patrick Head and another engineer John Piper. By December 1982 it was ready. What a car it was, with four-wheel drive, a mid-mounted V6 engine, fuel-injection and lightweight carbonfibre/Kevlar body. In tests, it proved itself far superior to its opposition. Development continued until 1986, and BL had good reason to be confident about its new car.

DISASTER STRIKES

Then came a bombshell. Following a series of high profile accidents, Group B cars were suddenly banned. The MG Metro 6R4 was now a rally car without a home. Such bad luck was typical for BL.

There was little choice for Austin Rover Motorsport. It gave up on rallying, and sold off most of its 6R4's into private hands. Just 200 homologation cars had been built, plus 20 even more savage Evolution models. This was not quite the end, though. The V6

engine, in modified form, later found its way into the Jaguar XJ220 supercar.

MG Metro 6R4

Top speed:	225km/h (140mph)
0–96km/h (0–60mph):	4.5 secs
Engine type:	V6
Displacement:	2991cc (182.5ci)
Max power:	186.5kW (250bhp) @ 7000rpm
Max torque:	305Nm (225lb-ft) @ 6500rpm
Weight:	1030kg (2266lbs)
Economy:	7.08km/l (20mpg)
Transmission:	Five-speed manual
Brakes:	Four-wheel vented discs
Body/chassis:	Chassis-less construction with multi-tubed underframe, suspension subframes and integral roll cage and carbonfibre and Kevlar body panels

UNITED KINGDOM

MORGAN PLUS 8

Morgans may look old-fashioned, but the marque's loyal customers refuse to have it any other way. And with potent V8 power, performance is anything but vintage.

The styling of the Morgan may look prewar, but it actually dates from the 1950s, and the Plus Four and 4/4 Morgans. There were some subtle differences, though, such as a slightly longer wheelbase. Since then, very little has been done to modify the looks.

Plus 8s use Rover V8 engines for power, as used in the Range Rover, MGB GT V8, SD1 and P6 sedans and countless specialist, low-volume sports cars. Since 1968, engine size and power output has varied, but Morgan has always stayed true to its Rover roots.

Fuel injection and front disc brakes are rare Morgan concessions to modernity.

The Morgan has excellent handling, thanks to the engine being set well back under the hood.

The hood lids on the Morgan open sideways.

Although Morgan has embraced some modern technology, suspension on the Plus 8 remains strictly traditional. At the front is a unique sliding pillar independent system, first used by Morgan in 1909, although it is now complimented by telescopic shock absorbers. At the rear – and almost as old-fashioned – is a live axle with leaf springs.

To those who do not understand the Morgan philosophy, it must seem amazing that this most anachronistic of British car companies has managed to survive into the twenty-first century. In a world where the newer and more technologically advanced that something is, the more highly it is regarded, Morgans must seem like outdated jokes.

However, the firm's long waiting lists tell a different story. Morgan ensures its continuing success simply by staying the same. Blending modern power and performance with traditional looks is what

Rover's V8 engine, which has always been used in the Morgan Plus 8, was originally a redundant American Buick design, purchased by the British company back in the 1960s.

Morgan has always done best. Why change the recipe now?

V8 VINTAGE

The Plus 8 was first seen in 1968, although 'first seen' is not quite the right phrase, as it looked just like the preceding Plus 4. Which in turn looked like something out of the 1930s. What was different was its engine, a Rover V8 3528cc (215ci), which gave a big jump in power and torque, and unleashed a lot of previously hidden potential. It may have looked different, but performance was definitely up to modern standards.

With a company like Morgan, there was no chance of the Plus 8 disappearing after a few years, and it remains in production today. The V8 has varied in power in the intervening decades, brakes have changed from drum to disc, a five-speed transmission came along in 1977, fuel-injection appeared in 1984, and there was a rack-and-pinion steering from 1986. Yet the Plus 8 is still resolutely old-fashioned, a typical Morgan and the only truly vintage car you can still buy new today. No doubt it will be around for decades to come.

Morgan Plus 8 1992 Model

Top speed:	195km/h (121mph)
0–96km/h (0–60mph):	6.3 secs
Engine type:	V8
Displacement:	3946cc (241ci)
Max power:	142kW (190bhp) @ 4750rpm
Max torque:	318Nm (235lb-ft) @ 2600rpm
Weight:	936kg (2059lbs)
Economy:	5.95km/l (16.8mpg)
Transmission:	Rover five-speed manual
Brakes:	Discs at front, drums at rear
Body/chassis:	Steel ladder frame with cross braces. Wooden frame and alloy or steel panelled two-door convertible body.

PANOZ ROADSTER

The extreme and exciting Panoz AIV Roadster from America forgot all the usual conventions of car design to focus on just two things: speed and handling.

A Ford V8 was used for power, and as the Roadster was all about performance, there was really only one choice of unit. It had to be a high specification Mustang unit, with 32 valves, and fuel injection fitted.

The body was built of aluminium, using aviation strengthening principles. The space frame chassis used the same material, based on a design by Frank Costin, whose previous credentials included Lotus, Maserati, Marcos and Lister.

Panoz raided disparate company parts bins to build the Roadster. As well as the Ford Mustang engine and five-speed manual transmission, the rear axle and differential were from Lincoln, the central brake light from the Ford Taurus, tail lamps from a Land Rover, rear-view mirrors from the Mazda Miata and sun visors from the BMW Z3.

The Panoz badge, found at the front and rear of the Roadster, was designed by founder Daniel E Panoz. It comprises a stylized swirling Japanese Yin-Yang and a shamrock in the centre (in tribute to the car's and the company's Irish origins). Its red, white and blue are the American colours.

In 1988, while living in Ireland, 26-year old Daniel Panoz went for a job at the local Thompson Motor Company. Instead, he found out the firm was going bankrupt, and ended up buying the rights to one of its chassis, designed by Frank Costin (the man who put the 'Cos' in Cosworth).

A year later in the United States, Panoz set up the Panoz Automotive Development Company, aiming to build his own sports car, with the Costin chassis as basis. Within a year, the first Roadster had appeared.

The wild and raw Roadster was a response to what European sports car manufacturers

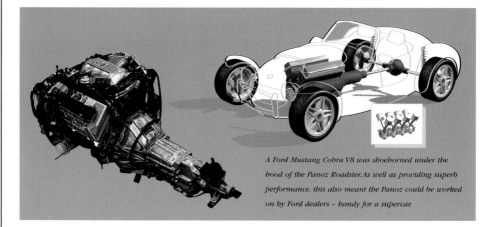

A Ford Mustang Cobra V8 was shoehorned under the hood of the Panoz Roadster. As well as providing superb performance, this also meant the Panoz could be worked on by Ford dealers – handy for a supercar.

were doing at the time. Electrifying in both performance and looks, such cars were also temperamental, prone to breaking down and expensive. By contrast, the Roadster was to be reliable, easy to maintain and competitively priced.

With any engine inside, the lightweight aluminium body and space frame chassis of the Roadster would have meant performance was great. However, with a Ford V8 under the bonnet, it was stupefying.

ALUMINIUM THROUGHOUT

For 1996, Panoz completely redesigned the Roadster while retaining its austere, stripped-out looks. This AIV (Aluminium Intensive Vehicle) version was now aluminium throughout, including the chassis and the Mustang V8. Panoz was the first US manufacturer to use the revolutionary AIV technique. It meant that performance was now simply unbelievable...

The Roadster just survived into the twenty-first century, being built until March 2000. Its successor was the more refined Esperante, a supercar that enlarged upon the engineering lessons principles of the Roadster.

Panoz Roadster

Top speed:	211km/h (131mph)
0–96km/h (0–60mph):	4.5 secs
Engine type:	V8
Displacement:	4600cc (281ci)
Max power:	227kW (305bhp) @ 5800rpm
Max torque:	406.5Nm (300lb-ft) @ 4800rpm
Weight:	1118kg (2459lbs)
Economy:	7.04km/l (19.9mpg)
Transmission:	Five-speed manual
Brakes:	Four-wheel vented discs
Body/chassis:	Aluminium space frame with two-door roadster body

UNITED STATES

257

PANTHER SOLO

The Solo was one of a kind, but its looks and performance could not save it from an ignominious end after just a handful had been built.

The original plans for the Solo involved a mid-mounted Ford Escort XR3i engine, but this was junked in favour of a Cosworth four-cylinder unit as used in the Cosworth Sierra sedan. Attached to the engine was an intercooled Garrett turbocharger.

It was the Panther's method of body construction that caused it to be so expensive. It was an intricate manufacturing process: a mixture of composite and fibreglass panels with two steel subframes holding the engine and suspension. The passengers were encased in an extremely strong 'survival cell', which passed crash testing with flying colours and also the 'unofficial' test of a 145km/h (90mph) accident on the A1 road in England.

The Panther's body was designed by Ken Greenly, an automotive design tutor at London's Royal College of Art, while the chassis was created by Len Bailey, who had worked on Ford's GT40.

The headlamps revolved instead of popping up, a feature that not only aided the streamlining but proved a major talking point as well.

The rear hatchback featured four large scoops to suck air into the rear-mounted window.

When the Panther Car Company launched its technologically advanced Solo 2 in 1987, it did so with the slogan, 'Only the few fly solo'. Unfortunately, this was spot on. Only a few people did buy a Solo, and its failure was enough to bring down the company.

Panther had been formed in 1972, and was known for building extravagant pastiches of 1930s-style classics. So when the Panther Solo 1 was revealed in 1984, people were surprised to see a good-looking sports car. The Solo was the vision of the Korean Yung Chull Kim, Panther's new chief executive.

Cosworth Sierra owners would have recognized the same engine inside the Panther Solo. The four-cylinder unit did not have a large displacement, but the Garrett turbocharger took care of any power deficiencies.

Then Kim went on holiday to Guam, where he tried out the new Toyota MR2. It was everything he had been trying to do with the Solo, and Kim telexed back to Britain for all work to stop.

So, it was back to the drawing board. What emerged from the ashes of the Solo 1 project was the Solo 2, which had a longer chassis, reworked aerodynamic body, four-wheel drive and more powerful, mid-mounted Cosworth engine.

SOLO LAID LOW

The new Solo debuted at the 1987 Frankfurt Motor Show, but Kim subsequently sold Panther for reasons that have never been fully explained. The new buyer was the (unfortunately named) Dong-A-Motor Company of Korea, who had little experience of building British supercars. Delays and financial problems set in, and once eager customers started cancelling their Solo 2 orders.

The plug was pulled on the Solo project in late 1990, after just 26 had been made.

Panther Solo

Top speed:	228.5km/h (142mph)
0–96km/h (0–60mph):	7.0 secs
Engine type:	In-line four
Displacement:	1993cc (122ci)
Max power:	152kW (204bhp) @ 6000rpm
Max torque:	268Nm (198lb-ft) @ 4500rpm
Weight:	1237kg (2723lbs)
Economy:	7.79km/l (22mpg)
Transmission:	Five-speed manual with four-wheel drive and centre and rear viscous couplings
Brakes:	Four-wheel vented discs, ABS standard
Body/chassis:	Steel floorpan and front and rear subframes with composite central cell and bodywork

PONTIAC FIREBIRD TRANS AM

Once every kid wanted a Trans Am because of TV's 'Knight Rider'. And adults wanted the Turbo because it was Pontiac's fastest 1980s production car.

The majority of the 20th anniversary models had targa roofs. Some of these cars were subsequently converted into convertibles by the American Sunroof Corporation, with the full approval of Pontiac.

The Turbo was gadget-laden inside, with electric windows, power steering, an adjustable steering wheel and cruise control. In addition, it had the controls for its stereo system located on the steering wheel.

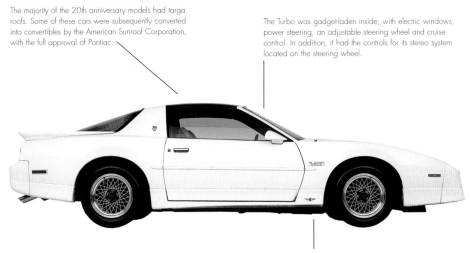

This badge on the fender commemorated the fact that the Turbo Trans Am had been the pace car at 1989's Indianapolis 5000.

This model is a 1989 Pontiac Turbo Trans Am twentieth anniversary model, released for one year only to celebrate two decades of one of America's most famous motoring names. Limited to 1555 models, it was available in white only, and came with a Garrett AiResearch turbocharger and intercooler bolted onto its V6 engine, plus a handling kit and sophisticated electronics to look after the mechanics.

Despite its staggering abilities, the back end of the Trans Am was quite basic, with a live rear axle.

UNITED STATES

Before actor David Hassellhoff became famous for running around in slow motion on beaches with a scantily clad Pamela Anderson in *Baywatch,* he was better known for hanging around with a black Trans Am. As curly permed, leather-jacketed Michael Knight, Hasselhoff battled crime with the aid of his car, KITT. But this was no ordinary Trans Am. Packed with all the latest electronics and possessing the ability to speak, plus a nice line in sarcasm, KITT was undoubtably the star of *Knight*

Going against V8 type, the 20th Anniversary edition of the Trans Am featured a V6 borrowed from Buick. However, the addition of a Garrett turbocharger meant it was a great performer.

Rider, one of the most popular TV shows of the 1980s.

Such exposure catapulted the third generation of Firebird Trans Am to instant worldwide fame soon after it was launched in 1982. However, this small screen glamour did a good job of disguising the fact that the Pontiac was just not terribly good. Its base model 'Iron Duke' version could manage only a paltry 67kW (90bhp) – not much of a supercar, then – and higher-spec versions could do little better.

PONTIAC PROGRESS

However, the Trans Am got better as the decade progressed. True supercar status belatedly arrived in 1989, when Pontiac launched its limited edition 20th anniversary Trans Am. It may have had only a V6, but it was fitted with a turbocharger, which allowed a top speed of over 241km/h (150mph). It was the most potent Pontiac of the 1980s.

The third-generation models ended in 1992, but check any satellite TV channel, and you'll be bound to find one forever in action in *Knight Rider.* The 1980s-era Trans Am will be winning fans for a long time to come.

Pontiac Firebird Trans Am

Top speed:	253km/h (157mph)
0–96 km/h (0–60 mph):	5.1 secs
Engine type:	V6
Displacement:	3785cc (231ci)
Max power:	190kW (255bhp) @ 4000rpm
Max torque:	461Nm (340lb-ft) @ 2800rpm
Weight:	1548kg (3406lbs)
Economy:	9.56km/l (27mpg)
Transmission:	GM 700R4 four-speed automatic
Brakes:	Four-wheel discs
Body/chassis:	Unitized frame with steel coupe body

UNITED STATES

PORSCHE 356

The story starts here. The 356 brought the Porsche name to international prominence, and paved the way for one of the world's greatest supercar marques.

Ferdinand Porsche had designed the Volkswagen Beetle before World War II, and the postwar 356, laid down by his son Ferry, continued with the same air-cooled, rear-mounted engine theme, as well as using torsion-bar suspension and swing axle rear suspension. It was a combination that could make the Porsche's handling delicate.

The Porsche 356 was available as a coupe, cabriolet and speedster. This example is a 1956 speedster, fitted with a 1582cc (96ci) engine.

A small circular cover in front of the rear wheels was an access hatch allowing the suspension torsion bar to be removed.

Early cars had light-alloy bodies, as the engines used were largely standard (and gutless) Volkswagen Beetle units, plus there was a shortage of steel due to the war. By 1950, steel-bodied cars had started to be built, and increases in engine capacity followed soon afterwards. Porsche could not cope with initial demand, so the Reutter firm of Stuttgart constructed the shells instead. These were not always put together well, and tended to rust badly.

Most 356s used drum brakes for stopping, although the 356C of 1963 did feature discs.

Like father, like son. It was Dr Ferdinand Porsche who had designed the iconic Volkswagen Beetle, but it was his son Ferry who would be responsible for coming up with the next great Porsche classic. And this time, it would bear the family name as well.

The 356th project to be undertaken by the Porsche design company since 1930 was a streamlined, simplistic sports coupe, with an exquisite teardrop shape. It looked gorgeous, even though underneath the heartstopping beautiful shape was just humble Beetle mechanicals.

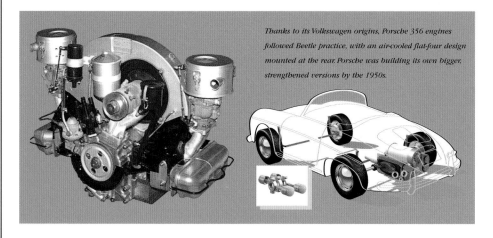

Thanks to its Volkswagen origins, Porsche 356 engines followed Beetle practice, with an air-cooled flat-four design mounted at the rear. Porsche was building its own bigger, strengthened versions by the 1950s.

Despite this, the Porsche 356 was an exciting car, and its aerodynamic shape meant that it was significantly faster than its VW ancestor. It was quick to find favour with private racers, as well as those who simply appreciated its European chic. The Hollywood star James Dean was one celebrity owner.

THE PORSCHE FORMULA

Demand was more than Porsche had anticipated, and as sales grew, so did the company and the 356. Larger engines were a logical move, and by 1955, the four-cylinder unit had grown to 1582cc (96.5ci). There was added sophistication too, the appeal of the range being boosted by the introduction of a cabriolet in 1951, and the Speedster of 1954.

A facelift introduced a revamped look for 1955, when the car became the 356A. The 356B of 1959 introduced a sloping bonnet, new bumpers and headlamps.

While not a supercar by today's standards, the 356 was a highly desirable car in the austere world of the 1950s. Without its success and template for the future, Porsche's later supercars might never have

Porsche 356

Top speed:	161km/h (100mph)
0–96km/h (0–60mph):	11.2 secs
Engine type:	Flat four
Displacement:	1582cc (96.5ci)
Max power:	52kW (70bhp) @ 4500rpm
Max torque:	111Nm (82lb-ft) @ 2700rpm
Weight:	813.5kg (1790lbs)
Economy:	9.2km/l (26mpg)
Transmission:	Four-speed manual
Brakes:	Four-wheel drums
Body/chassis:	Pressed steel platform chassis with steel bodywork, coupe or convertible

GERMANY

269

PORSCHE 911 TURBO 3.3 SE

The original air-cooled, rear-engined Porsche 911 is probably the world's most recognizable supercar,
with the Turbo SE version being one of its finest incarnations.

The first Turbos of 1975 had 3000cc (183ci) all-alloy engines, but by 1977, these had
gone up to 3300cc (201ci). Gas was fed directly by Bosch K-Jetronic fuel injection. A
single KKK turbocharger provided the extra boost, although the use of just one unit did lead
to turbo lag. This would later be resolved by the fitment of two smaller turbos.

Besides this main exhaust, there was a
smaller pipe at the rear, which acted as the
wastegate for the turbocharger.

For the SE, Porsche managed to squeeze an additional 22kW (30bhp) out of the existing Turbo engine. This was done by fitting better camshafts, a bigger turbo and intercooler, and a freer-flowing exhaust.

The huge whale-tail spoiler was the view most drivers got of a 911 Turbo!

This is a 1985 911 Turbo 3.3SE, the most exclusive Turbo version to be built. As well as engine modifications, the most obvious alterations were straked vents on the rear fenders for extra air to the engine, plus a sloping nose with pop-up headlamps, echoing Porsche's 935 racer.

GERMANY

271

Any Porsche 911 qualifies as a supercar, of course. But many enthusiasts, if forced to decide which particular version of the original air-cooled icon was the greatest, would opt for the mighty and often scary Turbo.

But there is one Turbo that stands head and shoulders - or perhaps that should be head and whale-tail spoiler - above the rest. In 1985, a decade after the Turbo's launch, Porsche cranked up the pressure on its rivals even more by letting loose the Turbo 3.3 SE.

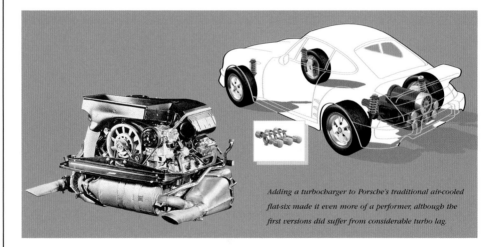

Adding a turbocharger to Porsche's traditional air-cooled flat-six made it even more of a performer, although the first versions did suffer from considerable turbo lag.

The initials stood for Sport Equipment, but this was more than just a relabelled 911 to which a few extras had been added in order to justify a higher price. At over twice the price of a standard Turbo, the SE did a lot to justify its expense.

For starters, there was 10 per cent extra power, which helped make the SE the fastest 911 ever built. Inside, the cabin featured such goodies as full leather trim, an electrically operated sunroof, air conditioning, heated and power-adjustable seats and a powerful stereo system.

GOING FLAT

But it was the exterior that really grabbed the eye. The whale-tail spoiler meant it looked just as flamboyant as an 'ordinary' Turbo, but the SE was also highly distinctive in its own right. It had a 'slant nose', with the fenders falling into line with the front hood. This meant dispensing with the exposed headlamps, so retractable ones were fitted instead. The resemblance to the 935 racer was unmistakable.

Production of this exclusive model was halted in 1987, although the flat nose became a high cost option on standard Turbos thereafter.

Porsche 911 Turbo 3.3 SE

Top speed:	274km/h (170mph)
0–96km/h (0–60mph):	5.0 secs
Engine type:	Flat six
Displacement:	3299cc (201ci)
Max power:	246kW (330bhp) @ 5750rpm
Max torque:	431Nm (318lb-ft) @ 4000rpm
Weight:	3000lbs (1364kg)
Economy:	6.37km/l (18mpg)
Transmission:	Four-speed manual
Brakes:	Four-wheel vented discs
Body/chassis:	Unitary monocoque construction with steel two-door coupe body

GERMANY

PORSCHE 928

Intended as a successor to the iconic 911, the 928 was Porsche's first purebred front-engined sports car, a departure from the firm's usual successful recipe.

This is not a standard 928. Several firms offered specially modified cars for customers with enough money to spend, and this one was prepared for a Porsche executive by Gemballa Coachworks. Among the departures from the norm are brakes to Formula 1 standard, extended and flared bodywork, and twin turbochargers fitted to the uprated engine. The most obvious change is the lack of a roof: Porsche never made any convertible 928s.

Although capable of an impressive top speed and fierce acceleration, the 928 was more grand tourer than out-and-out supercar. Consequently, most came with a four-speed automatic gearbox from Daimler Benz, although a five-speed manual was always available for those who liked a bit more control.

The emphasis was firmly on luxury in the 928, with plush upholstery, as well as a comprehensive range of equipment. The car was a 2+2, meaning there was accommodation for rear seat passengers, although the lack of space made it unsuitable for long trips!

A distinctive feature of the 928 was its circular headlamps, which retracted into the front fenders when not in use.

By the mid-1970s, the Porsche 911 was ageing. Its rear-engined, air-cooled design was perceived as outdated, and its handling was always on the wayward side, particularly if pushed hard. It seemed that some sort of replacement was called for. The trouble was, everybody adored the 911, and pinpointed its shortcomings as examples of its character.

So Porsche was presented with an extraordinarily tough job trying to come up with a potential successor. It decided to take the conventional route and build a front-engined, water-cooled supercar. This was

The front location of the 928's engine was conventional, which explains why Porsche enthusiasts failed to appreciate it as much as the 911, despite the fact that it was a far better performer.

not Porsche's first flirtation with this format – there had been the 924 collaboration with Volkswagen Audi – but the V8-engined 928 was total Porsche, and very different to the four-cylinder 924.

In looks, it was smooth and muscular, while in character, it was well-mannered and disciplined. Some critics called it docile, but its top speed of around 241km/h (150mph) refuted that claim. It was just an all-round competent package, with better stability than people were accustomed to expect from Porsche, and it was voted the International Car of the Year.

YEARS OF CHANGE
In 1979, the 928S, with a bigger, more powerful engine, joined the fold, and in 1986, the 928 S4 was launched with revised nose and tail sections and a 4957cc (305ci) engine. Porsche's policy of continual improvement subsequently resulted in the GT, and in 1992 came the GTS. This had a 5399cc (329ci) engine pumping out 261kW (350hp).

After 18 years in production, the 928 was dropped, while the 911, the car it was supposed to supersede, continued.

Porsche 928 Gemballa-Modified

Top speed:	264km/h (164mph)
0–96 km/h (0–60mph):	5.0 sec
Engine type:	V8
Displacement:	4957cc (302.5ci)
Max power:	335.5kW (450bhp) @ 6800rpm
Max torque:	515Nm (380lb-ft) @ 4400rpm
Weight:	1622kg (3568lbs)
Economy:	2.83km/l (8mpg)
Transmission:	Four-speed automatic
Brakes:	Four-wheel discs
Body/chassis:	Unitary monocoque construction with steel body

PORSCHE 911 RUF

If a standard Porsche 911 Turbo was not enough, the German Ruf tuning company offered an uprated version with four-wheel drive and twin turbochargers.

Considering the top speed of almost 322km/h (200mph), the optional extras of a full roll cage and harness were probably wise choices!

At first glance, the Ruf version of the 911 looked standard. However, the big giveaway that this was something special were the huge and very distinctive alloy wheels, with Ruf center caps, unique to the company. For better traction, the tyres were wider at the rear.

Power is nothing without control, and the Ruf got around the 911's tail-happy tendencies by employing four-wheel drive. It used the system from the Carrera 4, with a central viscous coupling and traction control.

As well as a conventional clutch, Ruf customers could opt for a clutchless electronic gearbox.

Ruf completely reworked Porsche's flat six engine, to turn it into a fire-breathing monster capable of forcing out almost 373kW (500bhp). The pistons were changed, and modified camshafts were fitted. And instead of the standard turbocharger setup, Ruf bolted on two enhanced KKK turbos. The final touches were to fit a less restrictive exhaust and rework the fuel management system.

GERMANY

279

The history of the Ruf tuning firm dates back almost as far as the company it would come to be most associated with. German national Alois Ruf started Auto Ruf in 1939 (Porsche was founded in 1931), and he began modifying Volkswagen Beetles after the war . But while Porsche transformed VWs into something more sporting (the 356 model), Ruf built economy versions of the People's Car.

From the 1960s onwards, Ruf started to get involved with Porsches, and modified its first

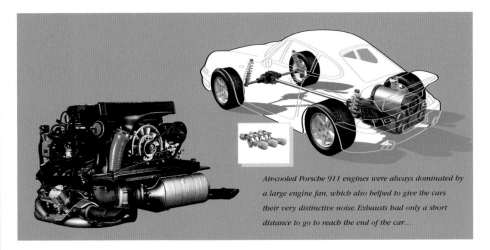

Air-cooled Porsche 911 engines were always dominated by a large engine fan, which also helped to give the cars their very distinctive noise. Exhausts had only a short distance to go to reach the end of the car...

911 Turbo in 1977. From there, it never looked back, building such ground-breaking models as the superb BTR and the shocking 'Yellow Bird' CTR, which broke the world speed record for production cars in 1987, with a speed of 339.5km/h (211mph).

TURBO R TERROR

When Porsche updated the 911's shape in 1993, this was a cue for Ruf to unleash one of its most monstrous models. The Turbo R was pure animal, its pugnacious spirit disguised by its smooth and sophisticated appearance. It looked very much like a stock 911, but even the shortest drive would reveal its true colours.

The Carrera 4, on which the Turbo R was based, was totally reworked. With twin turbochargers, four-wheel drive and a completely modified engine, Ruf's outlandish take on the 911 made standard Porsches look like...well, like the Beetles Ruf had once turned out years before.

Porsche launched its all-new water-cooled 911 in 1997, but Ruf refused to play along at first. It continued to offer its air-cooled Turbo R conversion as before. Some things just are difficult to improve on.

Porsche 911 Ruf

Top speed:	309km/h (192mph)
0–96km/h (0–60mph):	3.8 secs
Engine type:	Flat six
Displacement:	3600cc (220ci)
Max power:	365kW (490bhp) @ 5500rpm
Max torque:	650Nm (480lb-ft) @ 4800rpm
Weight:	1405kg (3090lbs)
Economy:	3.89km/l (11mpg)
Transmission:	Six-speed manual with electronic clutch option
Brakes:	Four-wheel discs
Body/chassis:	Unitary monocoque construction with steel two-door coupe body

GERMANY

281

RENAULT ALPINE A110

From humble beginnings, the French firm of Alpine grew into the country's leading sports car maker, with the A110 responsible for much of its success.

The four-cylinder engine was rear-mounted not mid-mounted, as it was positioned behind the back axle. This could make the handling 'interesting' under certain conditions, especially if the front fuel tank was getting low. All the various incarnations of the car used Renault sedan units, with this 1600S model being fitted with the overhead-valve engine from a R16.

Because it was primarily designed as a competition car, room for luggage was not really a consideration when it came to putting the A110 together. The engine took up the rear hood, while the boot at the front was dominated by the fuel tank.

The A110 relied on a single tube backbone chassis for its structural integrity, as the bodywork was of all fibreglass construction.

Although Alpine built the A110, it was generally referred to as ——— a Renault Alpine because it used engines, transmissions and running gear from the big French company.

Rear suspension on most of the A110s was by swing axles, but the final cars were fitted with double wishbones at the back. All models had double wishbone suspension at the front, though.

B orn into a family that owned a Renault dealership, Jean Redele had a destiny that was probably mapped out from an early age. It was always going to involve cars. When he was old enough, he started competing in motorsport using a distinctly un-racelike 4CV, but it did not take long for him to graduate to building his own cars, using Renault parts and fibreglass bodies.

When Redele also started to build cars for other people, he set up the Alpine company. Renault was sufficiently impressed that it

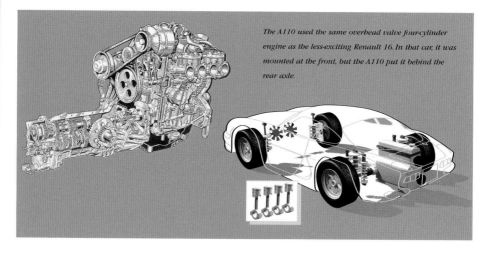

The A110 used the same overhead valve four-cylinder engine as the less-exciting Renault 16. In that car, it was mounted at the front, but the A110 put it behind the rear axle.

began selling them through its dealership network.

The definitive Alpine model arrived in 1962, when Redele unveiled the lightweight A110. It used Renault 8 and 8 Gordini parts, with the engine rear-mounted on a single tube backbone chassis. All of this was clothed in a fibreglass coupe body that had such timeless styling it would remain largely unchanged for the next 15 years.

RALLY AND ROAD

Intended first and foremost for rallying – something it excelled at – the austere Alpine made a very rough and ready road car, although one with compelling performance once the driver had mastered its tail-happy handling. Alpine did build a more civilized four-seater A110 GT4 from 1963 to 1965 purely for street use, but this failed to make much of an impression.

When Alpine fitted the more powerful engine from the 1605cc (98ci) Renault 16,

the A110's rallying success really took off. It dominated the world, winning in any sort of climate, on any road surface – truly a master of all events. Production ended in 1977.

Renault Alpine A110

Top speed:	204km/h (127mph)
0–96km/h (0–60mph):	6.3 secs
Engine type:	In-line four
Displacement:	1605cc (98ci)
Max power:	103kW (138bhp) @ 6000rpm
Max torque:	144Nm (106lb-ft) @ 5000rpm
Weight:	712kg (1566lbs)
Economy:	10.68km/l (23.5mpg)
Transmission:	Four or five-speed manual
Brakes:	Four-wheel discs
Body/chassis:	Single tube backbone chassis with separate fibreglass two-door coupe body

FRANCE

RENAULT ALPINE A610

Renault's purchase of the Alpine company led to more sophistication in the Alpine range, culminating in the GTA and A610 models of the 1980s and 1990s.

The A610 continued the traditional Alpine practice of installing the engine at the rear of the car, behind the rear axle. The fuel-injected all-alloy V6 featured an intercooled Garrett T3 turbocharger to boost power. Along with the rear double wishbone suspension, it was mounted on a separate subframe, so it could be completely removed if major engine work was necessary.

To balance out the engine at the rear, the fuel tank and the spare tyre were installed in the front boot. Unfortunately, this left hardly any space for luggage.

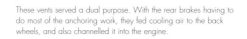

These vents served a dual purpose. With the rear brakes having to do most of the anchoring work, they fed cooling air to the back wheels, and also channelled it into the engine.

The body for the A610 was made of fibreglass, so corrosion was not an issue. It was bonded to a steel backbone chassis.

The previous incarnation of the A610 was the V6GT. When the A610 was created in 1991, the most noticeable change was the fitment of pop-up headlamps, in place of the glass-covered items of the V6GT.

Name a rear-engined six-cylinder European supercar with superb performance, exhilarating handling and a sleek, classic look so visually spot-on that it remained virtually the same for years.

The obvious answer is the Porsche 911, but in the 1980s and 1990s, Porsche's enduring icon had a similarly engineered rival to contend with. When Renault launched its V6GT in 1984, the Stuttgart supercar was its main opponent.

Renault bought the specialist Alpine company in 1974, and started building more advanced Alpines than previously, although

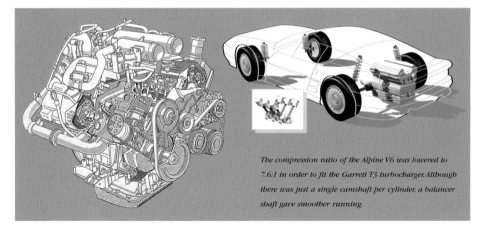

The compression ratio of the Alpine V6 was lowered to 7.6:1 in order to fit the Garrett T3 turbocharger. Although there was just a single camshaft per cylinder, a balancer shaft gave smoother running.

continuing to use the rear-wheel drive format. In 1984, it unveiled the smooth and stylish GTV6 (known as the GTA in some countries), powered by the uncomplicated but reliable alloy V6 engine developed jointly with Peugeot and Volvo.

PORSCHE WANNABE

In 1991, this metamorphosed into the more refined A610, using the same basic exterior and mechanical layout. To make it more cosmetically attractive in terms of 1990s styling, there were a few tweaks (including retractable headlamps), and the engineering was improved to make handling more compliant. Power got a big boost thanks to a single turbocharger, while ABS was fitted to make sudden braking a less heart-stopping process. Power steering also ensured that the A610 was less of a handful.

In many ways, the A610 was a more exciting car than the 911, and certainly an easier drive. But it was much less successful.

The chief reason was that it sported not a Porsche badge, but the less illustrious Renault one. In the fashion-conscious world of supercars, such things are often more important than what a car actually does.

Renault Alpine A610

Top speed:	256km/h (159mph)
0–96km/h (0–60mph):	5.9 secs
Engine type:	V6
Displacement:	2975cc (181.5ci)
Max power:	186kW (250bhp) @ 5750rpm
Max torque:	350Nm (258lb-ft) @ 2900rpm
Weight:	1383kg (3043lbs)
Economy:	6.41km/l (18.1mpg)
Transmission:	Five-speed manual
Brakes:	Four-wheel vented discs
Body/chassis:	Steel backbone chassis with rear subframe and bonded-on fibreglass hatchback coupe body

FRANCE

RENAULT SPORT SPIDER

Renault's Sport Spider was a raw and primitive sports car that refused to let anything as frivolous as comfort get in the way of performance.

There was practically no weather protection for Spider passengers. The first versions did not even have windscreens, just a wind deflector. When Renault started building right-hand drive versions, it relented slightly and installed a windshield and wiper, plus an emergency soft-top roof that was no good at speeds of over 64km/h (40mph)!

No steel was used in either the Spider's body or its chassis. The body was constructed out of fibreglass, while the chassis was an aluminium spaceframe.

The interior was completely stripped out, with only the figure-hugging Recaro seats and the bare essentials necessary to drive the Spider inside the cockpit. In front of the driver was a large rev counter, while the digital speedometer was almost a secondary instrument, off to the side.

Power came from a fuel-injected, twin-cam, four-cylinder, 16-valve engine as used in the Megane coupe. It was mid-mounted in the Spider, though. Because the car was so light, the moderate 112kW (150bhp) output was still enough to endow the Spider with considerable performance, although the well-built chassis was capable of handling a lot more.

FRANCE

From any company, the back-to-basics Sport Spider would have been an unusual surprise. From a mainstream automotive maker like Renault, more used to sensible cars than sensational ones, it was nothing short of a revelation.

At the start of the 1990s, the visionary Patrick Le Quement took over as the design head of Renault. With his appointment, old ways of thinking were swept away, leading to such bold creations as the Twingo, Vel Satis and the awe-inspiring Avantime.

You could also have found the Spider's 16-valve, in-line four engine in the then current Clio Williams and Megane 16-valve Coupe – but not in a mid-mounted position, and not with quite the same level of performance.

But his most dynamic offering was the 1995 Renault Sport Spider. There was no nonsense about this car, it was nothing less than a racing car made road legal. Built with a mid-mounted engine and an incredibly lightweight aluminium chassis, it got rid of everything that was not absolutely essential. The first cars had no front screen, just a wind deflector intended to send air over the top of anyone inside – Renault recommended the wearing of helmets! For obvious reasons, this did not really work, so Renault installed a conventional screen the following year.

RACING ROADSTER

On road and track, it was unshakeable, with handling like a slot car. Corners were treated just as gentle curves, the Sport Spider gliding over the surface. Renault even started its own one-make racing series to demonstrate what the car was capable of.

In sales terms, the Spider was a failure, production ending in 2000, after only small numbers had been made. In publicity terms, though, its effect was enormous, helping to reinvent Renault as a manufacturer perceived to be both brave and unpredictable.

Renault Sport Spider

Top speed:	199.5km/h (124mph)
0–96km/h (0–60mph):	7.7 secs
Engine type:	In-line four
Displacement:	1998cc (122ci)
Max power:	112kW (150bhp) @ 6000rpm
Max torque:	190Nm (140lb-ft) @ 4500rpm
Weight:	957kg (2106lbs)
Economy:	10.68km/l (23.5mpg)
Transmission:	Five-speed manual
Brakes:	Four-wheel vented discs
Body/chassis:	Extruded and welded alloy frame with fibreglass, two-door, two-seat convertible body

FRANCE

SWALLOW SIDECAR SS100 (JAGUAR)

The SS100 played a major role in establishing the sporting credentials of SS Cars. It also saw the first use of a great name: Jaguar.

The distinctive SS100 shape was born out of the preceding SS90 model. It also set a precedent for future Jaguar styling, with its long bonnet, short tail and flowing, low-slung lines. Later XK series cars, the E-type and even current Jaguar sports cars all displayed echoes of the SS100.

The body was a mix of steel and alloy panels laid over an ash wood frame.

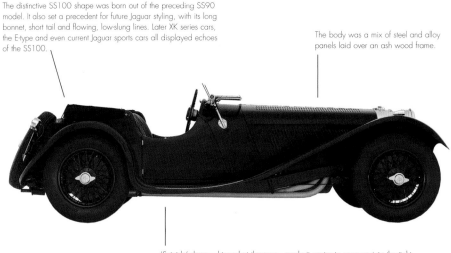

'Suicide' doors – hinged at the rear – made it easier to squeeze into the tight cabin. Their big disadvantage was that they could sometimes fly open without warning, particularly in convertible cars which suffered from body flexing.

The large windscreen could be folded down, and two tiny 'aero-shields' used instead, to improve the aerodynamics. Such was the exquisite attention to detail on the SS100 that the windscreen knob carried the SS monogram.

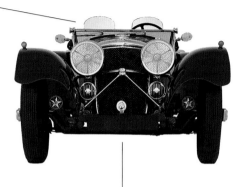

There was no external boot lid, but some luggage could be fitted behind the seats from inside the car.

The 2663cc (162.5ci) and 3485cc (213ci) engines were based on Standard Motor Company sidevalve units, but modified by tuning expert Harry Weslake with overhead valve cylinder heads.

UNITED KINGDOM

It was the prewar SS100 which began Jaguar's reputation as a manufacturer of sports cars par excellence, and paved the way for future glories.

The SS100 was one of the most beautiful cars of its era, with overtones of Italian and German flair in its styling. Yet it was the work of somebody very British indeed, William Lyons, boss of SS Cars. Lyons fledgling marque had started out building motorcycle sidecars, and then moved into special bodies for Austin Sevens. However, it was the SS100 that made the general public begin to take real notice of Lyons and SS.

The SS100's motor, designed by vintage tuning legend Harry Weslake, was the first overhead valve engine to be used by the firm, and employed a robust seven-bearing crankshaft.

Its handsome, rakish lines were as good-looking as any of its contemporaries, but its price tag was far more affordable. Lyons' business acumen, coupled with the use of mechanical parts from the Standard Motor Company, meant that it was almost ridiculously cheap. However, Lyons brought in the freelance tuning expert Harry Weslake to endow the Standard engine with more performance. He developed an overhead valve head that could be fitted on the existing block. This meant that the SS100 could live up to its title by being (almost) capable of 160km (100mph).

JAGUAR ARRIVES

The car had another designation, though. Seeking a further label that would personify the character of SS Cars, Lyons also bestowed the Jaguar title on the range. It was the birth of a famous name.

The war curtailed production after just five years, but not before the engine had been enlarged from 2663cc (162.5ci) to 3485cc (213ci) in 1938. All SS100s were convertibles, except for a solitary 1938 coupe show car.

Swallow Sidecar SS100

Top speed:	153km/h (95mph)
0–96km/h (0–60mph):	13.8 secs
Engine type:	In-line six
Displacement:	2663cc (162.5ci)
Max power:	77.5kW (104bhp) @ 4500rpm
Max torque:	219.5Nm (162lb-ft) @ 2750rpm
Weight:	1170kg (2575lbs)
Economy:	4.96km/l (14mpg)
Transmission:	Four-speed manual
Brakes:	Rod-operated drums
Body/chassis:	Two-seater sports car body on pressed-steel ladder chassis

UNITED KINGDOM

TVR Cerbera

With its name recalling a mythic beast from Hades, the Cerbera is a monster car, a rough 'n tough beast with smooth looks and brutal strength.

TVR's origins as a low volume specialist car manufacturer meant it usually bought in engines from mainstream British manufacturers like Ford and Rover. The Cerbera, though, was the first road car to use TVR's new AJP V8 unit, designed by Al Melling especially for the firm. Initially, it was used for racing purposes.

There are no handles on the doors, and the way of opening them is inventive. Both exterior mirrors have a small button on them. Press it, and the door is unlocked and the electric window lowered. Not only does this make the body smoother, but it is also bound to impress anybody you take along as a passenger!

Breaking with tradition, the Cerbera is a 2+2 instead of a genuine two-seater like its recent relations. Space in the back is limited, though. A suitcase or set of golf clubs is more likely to be comfortable than an average-sized adult!

TVR has always built fibreglass-bodied cars, and the Cerbera stays true to the faith. However, it is still an advanced design, with the roll cage bonded into the shell.

According to ancient mythology, Cerberus is a three-headed dog, the guardian of the gates of Hell. And according to more modern myth, a dog was involved in the design of the Cerbera. The fable goes that the cut-outs around the front indicators were created when Ned, the pet dog of the TVR chairman, sunk his teeth into a small-scale styling model of the proposed new car. Everybody liked the feature, and so the happy accident was left as it was.

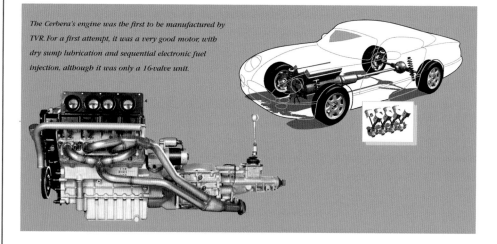

The Cerbera's engine was the first to be manufactured by TVR. For a first attempt, it was a very good motor, with dry sump lubrication and sequential electronic fuel injection, although it was only a 16-valve unit.

If the story is true, Ned is probably the first canine in history to have co-designed a supercar. And even if the story is false, it remains a terrific tale.

DOG STAR

The Cerbera first appeared in 1993, and then again in 1994 and 1995. These motor show versions were not available to the public, but certainly kept appetites whetted for when the car finally did go on sale in 1996.

And was the wait worth it? Absolutely. The TVR Cerbera was simply superb. It looked amazing, an aerodynamic four-wheeled projectile with perfect poise and arrogant attitude. Underneath this fibreglass sculpture of speed was a strong multi-tubular steel chassis which endowed the car with awesome handling. And growling away ferociously under the hood was a very special AJP V8 engine, the first TVR had ever commissioned for itself.

Even faster than a Dodge Viper (a car with an engine almost twice the size of the TVR's), the Cerbera continues to be built today. Ned the dog has every reason to be proud of himself.

TVR Cerbera

Top speed:	270km/h (168mph)
0–96km/h (0–60mph):	4.1 secs
Engine type:	V8
Displacement:	4475cc (273ci)
Max power:	313kW (420bhp) @ 6750rpm
Max torque:	515Nm (380lb-ft) @ 5500rpm
Weight:	1181kg (2598lbs)
Economy:	6.83km/l (19.3mpg)
Transmission:	Five-speed manual
Brakes:	Four-wheel vented discs
Body/chassis:	Separate chassis with two-door coupe body with fibreglass panels

UNITED KINGDOM

ULTIMA SPYDER

Faster than Ferrari, Porsche, Lamborghini and almost anything else, an Ultima could be beaten only by a McLaren F1. And no-one could self-build the McLaren at home...

The entire rear wing can be removed in seconds, either for maintenance on the mid-mounted engine, or so that a differently shaped rear can be fitted.

As this is a kit car, it is up to the owner to fit whatever engine he wants. The firm's preference is a small-block Chevrolet V8, developing between 224kW (300bhp) or an awesome 447kW (600bhp), but Rover V8s and Renault V6s have also been used.

The cockpit may look tight, but it is able to take occupants well over 1.8m (6ft) tall. On right-hand drive cars, the gear lever is mounted to the right of the steering wheel if a Porsche gearbox is fitted, or to the left if a Getrag gearbox is installed.

A CD-ROM version of the manual is sent out when an Ultima is ordered. The hard copy comes only when the kit itself arrives. No specialist tools are required, and the company claims that 75 per cent of buyers are first-time builders.

The tubular steel chassis is intensely strong, capable of taking 746kW (1000bhp) without stress.

Kit cars rarely fit into the supercar category. Often cheap and cheerful, and using parts scrounged from nondescript mass-produced saloons, many of these home-built jalopies look both primitive and awkwardly styled.

There are exceptions to any rule, though, and the Ultima Spyder is one hell of an exception. Taking the shape of endurance racers – cars such as those seen at Le Mans – as its inspiration, it not only looks far more exotic than its price tag suggests, but it also

The choice of engines is up to the customer, but if you fit anything less than a Chevrolet V8, you are not being fair either to the Ultima Spyder or yourself. The lightweight sports convertible is a natural home for this slice of American beef.

has a speed and acceleration to surpass many supercars. In fact, with the ability to reach 160km/h (100mph) in just 6.8 seconds, it has the sort of performance most supercar manufacturers dream of.

The Ultima Sports – from which the Spyder is derived – was conceived by specialist manufacturer Lee Noble in the late 1980s. It was initially intended just for racing, but a road-legal version arrived in 1992. A year later came the open-top Spyder, based on the same chassis as the Sport. It opened up whole new markets for Ultima.

DIY DYNAMITE

Supplied either as a kit or factory-built, the Spyder has fibreglass or Kevlar/carbonfibre bodywork mated to a tubular steel chassis. Engines which can be fitted range from a Renault V6 to a monstrous Chevy V8, while the transmission is generally Porsche or Getrag. Minor components from other marques are also used.

The Spyder is still being sold, offering levels of driving exhilaration and supercar performance previously unheard of for this kind of cost.

Ultima Spyder

Top speed:	273.5km/h (170mph)
0–96km/h (0–60mph):	3.8 secs
Engine type:	V8
Displacement:	5733cc (350ci)
Max power:	257kW (345bhp) @ 5600rpm
Max torque:	513.5Nm (379lb-ft) @ 3600rpm
Weight:	363kg (799lb)
Economy:	6.37km/l (18mpg)
Transmission:	Porsche five-speed transaxle
Brakes:	Four-wheel AP racing vented discs
Body/chassis:	Separate tubular-steel chassis with fibreglass or Kevlar/carbonfibre bodywork

UNITED KINGDOM

VECTOR W8-M12

Flamboyant in both looks and performance, the Vector was America's supercar answer to Lamborghini. It was a fast and dramatic machine, but certainly not subtle.

The W8/WX3 used a Corvette 350 V8 engine, which produced some phenomenal outputs. With fuel injection and twin turbochargers, it was capable of 820kW (1100bhp), a mind-boggling level of power. When the later M12 switched to a Lamborghini V12, it could not even come close...

The razor-edged wedge shape echoed the Lamborghini Countach, which was unveiled around the same time as the prototype Vector was first seen at the 1972 Los Angeles Motor Show. This example is a W8, in production from 1990 to 1992, after which it was restyled.

The body panels were a mixture of Kevlar, fibreglass and carbonfibre, over a bonded aluminium and tubular steel honeycomb frame. For its time, this was a very advanced construction process; other supercar manufacturers did not start using composites for years.

306

Like the Lamborghini Countach, Vectors also had doors that opened by swinging upwards.

The interior was as spectacular as the exterior, with a hi-tech dashboard and digital displays enveloping the occupants.

The front boot was so low that the radiator was installed flat at the front. There was not much room for luggage either!

UNITED STATES

When asked what had inspired America's most extraordinary supercar, Gerald Wiegert, Vector's founder, said: 'I wanted to be a fighter pilot, but my eyesight's not good enough. So I decided to build a fighter plane for the street.'

Vector was founded in the early 1970s so that Wiegert could create his ultimate all-American supercar, something that would surpass even anything from Italy. A show car – a theatrical mass of straight edges, flat surfaces and angular lines – was displayed in

The W8/WX3 used a turbocharged version of Chevrolet's Corvette V8 350 for power, while the M12 had a Lamborghini Diablo V12. Incredibly, the US engine, with four fewer cylinders, was much more powerful than the advanced Italian offering.

1972, followed by the W2 in 1977, which was billed as 'the fastest car in the world'.

ROAD FIGHTER

And then it all went quiet, until 1990, and the W8. At last, Vector had a production car it could sell. Its implacable wedge design was both startling and formidable, while its tuned Corvette engine provided enormous potency. When twin turbochargers were added to 1992's more curvaceously restyled WX3, the car almost became the fighter plane Wiegert had always wanted, with an alleged top speed of 402km/h (250mph). Nobody was ever brave enough to try and find out if this was true, though.

Power struggles within Vector led to Wiegert leaving in 1993, and Lamborghini owner Megatech taking over. It put drastic plans in place to try and make the Vector a practical, financially viable proposition. It was relaunched in 1995, as the M12, with a Lamborghini Diablo V12 engine.

Unfortunately, the plan failed. The company went bust, and the M12 disappeared in 1997. In the quarter of a century since the first Vector, only around 20 cars had been manufactured.

Vector W8-M12

Top speed:	314km/h (195mph)
0–96km/h (0–60mph):	4.1 secs
Engine type:	V12
Displacement:	5707cc (348ci)
Max power:	367kW (492bhp) @ 7000rpm
Max torque:	580Nm (428lb-ft) @ 5200rpm
Weight:	1504kg (3308lbs)
Economy:	2.97km/l (8.4mpg)
Transmission:	Three-speed automatic
Brakes:	Four-wheel discs
Body/chassis:	Semi-monocoque honeycomb chassis with two-door coupe body in composite materials

UNITED STATES

VENTURI ATLANTIQUE

Compared to its neighbour Italy, France has a history of supercars that is far less varied. However, one Gallic firm building sports cars to rival the best is Venturi.

The Atlantique uses a plastic composite body, mounted on a separate backbone steel chassis.

These distinctive door mirrors could originally be found on the Citroën CX. Because of their unusual and air-smoothed shape, they have often appeared on specialist sports cars, such as the Lotus Esprit.

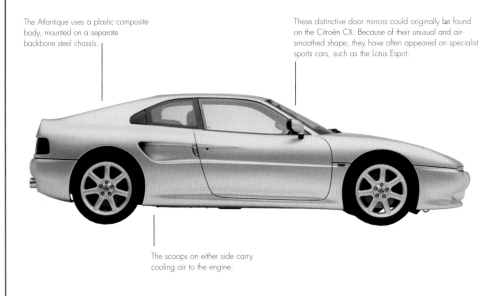

The scoops on either side carry cooling air to the engine.

The original Venturi was the brainchild of two French designers, Gerard Godfroy and Claude Poiraud, who both worked for the design company Heuliez. To develop the complicated suspension, they brought in Jean Rondeau who, as a racing driver, knew a thing or two about making cars handle well. He was tragically killed in 1985, though, so Mauro Bianchi and Jean-Pierre Beltoise – again, both racers – were left to continue what he had started. The current Atlantique has wishbones at the front, and multi-link suspension at the rear.

In best supercar tradition, the engine is mid-mounted. Developed by Renault and Peugeot/Citroën, it is a transverse V6 unit, made out of aluminium and featuring 24-valves, four overhead camshafts and fuel injection. Also fitted (on the Biturbo) are twin intercooled turbochargers.

FRANCE

Cars are an accurate reflection of national identities. The Italians build cars known for their temperamental passion, the Germans make vehicles that are reliable and efficient, and the best British cars have a touch of class, as well as a dash of eccentricity.

And France? Well, led by Citroën, and ably supported by Renault and Peugeot, French cars have always tried to be different. Others can do fast and conventionally stylish, France does quirky.

For that reason, there have been only a few French supercar manufacturers. Venturi is

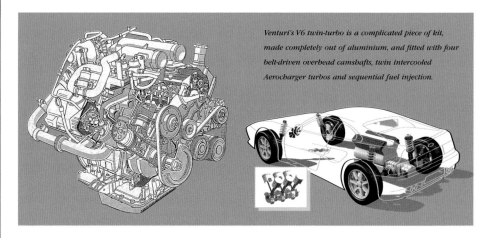

Venturi's V6 twin-turbo is a complicated piece of kit, made completely out of aluminium, and fitted with four belt-driven overhead camshafts, twin intercooled Aerocharger turbos and sequential fuel injection.

the exception, building the Atlantique, a composite plastic-bodied supercar in the mould of Ferrari and Lotus. Though out of the national character, it is a wonderful car.

FRENCH FANCY OR FOLLY?

The Atlantique 300 came out in 1994, nine years after the company was formed. Shown at the Paris Motor Show, it had been designed in just six months by co-founder Gerard Godfroy. Both the mid-mounted V6 engine and five-speed manual transmission were Peugeot/Citroën items, and, as well as the standard version, the Atlantique also came with a turbocharger. In this form, it generated approximately 300bhp (224kW), thus giving the 300 tacked onto the end of the Atlantique's name.

Its manufacture was low volume, just 250 of these expensive and exclusive sports cars a year being planned. In 1996, the original company went bankrupt, and a Thai consortium took over. It introduced the 300

Biturbo, with two small Aerodyne Dallas turbochargers. But in 2000, Venturi went belly-up yet again.

It now has a new owner, and production of this troubled supercar continues, albeit in tiny numbers.

Venturi Atlantique

Top speed:	280km/h (174mph)
0–96km/h (0–60mph):	5.3 secs
Engine type:	V6
Displacement:	180ci (2946cc)
Max power:	225kW (302bhp) @ 5500rpm
Max torque:	404Nm (298lb-ft) @ 2500rpm
Weight:	1250kg (2750lbs)
Economy:	8.50km/l (24mpg)
Transmission:	Five-speed manual
Brakes:	Four-wheel vented discs
Body/chassis:	Separate backbone chassis with composite two-door coupe body

FRANCE

313

INDEX